STOCHASTIC APPROXIMATION AND NONLINEAR REGRESSION

STOCHASTIC APPROXIMATION AND NONLINEAR REGRESSION

ARTHUR E. ALBERT
LELAND A. GARDNER, JR.

RESEARCH MONOGRAPH NO. 42
THE M.I.T. PRESS, CAMBRIDGE, MASSACHUSETTS

Set in Times New Roman
Printed and bound in the United States of America by
The Riverside Press, Cambridge, Massachusetts

Library of Congress catalog card number: 67–16501

To Lise

To Margie

Foreword

This is the forty-second volume in the M.I.T. Research Monograph Series published by the M.I.T. Press. The objective of this series is to contribute to the professional literature a number of significant pieces of research, larger in scope than journal articles but normally less ambitious than finished books. We believe that such studies deserve a wider circulation than can be accomplished by informal channels, and we hope that this form of publication will make them readily accessible to research organizations, libraries, and independent workers.

HOWARD W. JOHNSON

Preface

This monograph addresses the problem of "real-time" curve fitting in the presence of noise, from the computational and statistical viewpoints. Specifically, we examine the problem of nonlinear regression where observations $\{Y_n : n = 1, 2, \cdots\}$ are made on a time series whose mean-value function $\{F_n(\boldsymbol{\theta})\}$ is known except for a finite number of parameters $(\theta_1, \theta_2, \cdots, \theta_p) = \boldsymbol{\theta}'$. We want to estimate this parameter. In contrast to the traditional formulation, we imagine the data arriving in temporal succession. We require that the estimation be carried out in real time so that, at each instant, the parameter estimate fully reflects all of the currently available data.

The conventional methods of least-squares and maximum-likelihood estimation, although computationally feasible in cases where a single estimate is to be computed after the data have been accumulated, are inapplicable in such a situation. The systems of normal equations that must be solved in order to produce these estimators are generally so complex that it is impractical to try to solve them again and again as each new datum arrives (especially if the rate of data collection is high). Consequently, we are led to consider estimators of the "differential correction" type. Such estimators are defined recursively. The $(n + 1)$st estimate (based on the first n observations) is defined in terms of the nth by an equation of the form

$$\mathbf{t}_{n+1} = \mathbf{t}_n + \mathbf{a}_n[Y_n - F_n(\mathbf{t}_n)] \qquad (\mathbf{t}_1 \text{ arbitrary}; n = 1, 2, \cdots),$$

where $\{\mathbf{a}_n\}$ is a suitably chosen sequence of "smoothing" vectors. The term "differential correction" refers to the proportionality of the difference between $\mathbf{t}_{n+1}$ and $\mathbf{t}_n$ (the correction) to the difference between the nth observation, Y_n, and the value that would be predicted by the regression function if $\mathbf{t}_n$ were in fact the "true" parameter value.

The choice of smoothing vectors critically affects the computational simplicity and statistical properties of such recursive estimates. The main purpose of this monograph is to relate the large-sample statistical behavior of said estimates (consistency, rate of convergence, large-sample distribution theory, asymptotic efficiency) to the properties of the regression function and the choice of smoothing vectors. A wide class of smoothing vectors is examined. Some are deterministic and some depend on (are functions of) the observations.

The techniques used in the analysis are, for the most part, elementary and, by now, standard to those who are familiar with the literature of stochastic approximation. However, for the sake of the nonspecialist, we have tried to keep our treatment self-contained. In all cases, we seek the asymptotic properties (large n) of the solution to the nonlinear difference equation which relates $\mathbf{t}_{n+1}$ to $\mathbf{t}_n$.

As a fortuitous by-product, the results of this monograph also serve to extend and complement many of the results in the stochastic-approximation literature.

The structure of the monograph is as follows. Part I deals with the special case of a scalar parameter. Here we discuss probability-one and mean-square convergence and asymptotic distribution theory of the estimators for various choices of the smoothing sequence $\{a_n\}$. Part II deals with the probability-one and mean-square convergence of the estimators in the vector case for various choices of smoothing vectors $\{\mathbf{a}_n\}$. Examples are liberally sprinkled throughout the book. In fact, an entire chapter is devoted to the discussion of examples at varying levels of generality.

The book is written at the first-year graduate level, although this level of maturity is not required uniformly. Certainly the reader should understand the concept of a limit both in the deterministic and probabilistic senses. This much will assure a comfortable journey through Chapters 2 and 3. Chapters 4 and 5 require acquaintance with the Central Limit Theorem. Familiarity with the standard techniques of large-sample theory will also prove useful but is not essential. Chapters 6 and 7 are couched in the language of matrix algebra, but none of the "classical" results used are deep. The reader who appreciates the elementary properties of eigenvalues, eigenvectors, and matrix norms will feel at home.

The authors wish to express their gratitude to Nyles Barnert, who collaborated in the proofs of Theorems 6.1 through 6.3; to Sue M. McKay, Ruth Johnson, and Valerie Ondrejka, who shared the chore of typing the original manuscript; to the ARCON Corporation, the M.I.T. Lincoln Laboratory, the Office of Naval Research, and the U.S. Air Force Systems Command, who contributed to the authors' support during the writing of the monograph; and, finally, to the editorial staff of the *Annals of Mathematical Statistics*, who were principally responsible for the writing of this monograph.

ARTHUR E. ALBERT
LELAND A. GARDNER, JR.

Cambridge, Massachusetts
October 1966

Contents

1. Introduction

Despite the many significant and elegant theoretical developments of the past several decades, the art of statistical inference on time series is, from the applied point of view, in its infancy. An important class of problems, which has been relatively neglected, arises from the fact that there are always computations associated with statistical procedures; a procedure which is "optimal" in the decision theoretic sense can be somewhat less than optimal from a practical point of view if the associated computations are prohibitively lengthy. This dilemma is compounded when we consider a time series as a *flow* of data. In "space age" applications, it is especially important that statistical procedures keep pace with the incoming data so that, at any instant, all of the available information has already been processed. The acquisition of new observations merely serves to update the current state of knowledge.

In this monograph we will investigate nonlinear regression from that point of view. Let

$$\{Y_n : n = 1, 2, \cdots\}$$

be a stochastic process whose mean-value sequence is a member of a family of known sequences, that is to say,

$$\mathscr{E} Y_n = F_n(\boldsymbol{\theta}),$$

where $\boldsymbol{\theta}$ is a vector parameter which is not known and must be estimated.

We will explore the asymptotic (large n) properties of recursive estimation schemes for $\boldsymbol{\theta}$ of the form

$$\mathbf{t}_{n+1} = \mathbf{t}_n + \mathbf{a}_n[Y_n - F_n(\mathbf{t}_n)], \tag{1.1}$$

where $\mathbf{t}_{n+1}$ is the estimate of $\boldsymbol{\theta}$ based upon the first n observations and $\{\mathbf{a}_n\}$ is a suitably chosen sequence of "smoothing vectors."

Without question, estimators of the type of Equation 1.1 are computationally appealing, provided the smoothing sequence is chosen reasonably. After each observation, we compute the prediction error $Y_n - F_n(\mathbf{t}_n)$ and correct $\mathbf{t}_n$ by adding to it the vector $[Y_n - F_n(\mathbf{t}_n)]\mathbf{a}_n$. Such recursions are sometimes called "differential correction" procedures.

In contrast, maximum-likelihood and least-squares estimation methods, although often efficient in the purely statistical sense, require the solution of systems of simultaneous nonlinear normal equations. If we want "running" values of these estimates, the computational problems are often great.

Of course, the choice of the weights $\mathbf{a}_n$ critically affects the computational simplicity and statistical properties of the recursive estimate (Equation 1.1). The main purpose of this monograph is to relate the large-sample statistical behavior of the estimates to the properties of the regression function and the choice of smoothing vectors.

Estimation schemes of the type of Equation 1.1 find their origins in Newton's method for finding the root of a nonlinear function. Suppose that $G(\cdot)$ is a monotone differentiable function of a real variable, and we wish to find the root θ of the equation

$$G(x) = 0.$$

If t_1 were known to be a reasonably good estimate of (i.e., is close to) θ, then

$$0 = G(\theta) \approx G(t_1) + (\theta - t_1)\dot{G}(t_1), \tag{1.2}$$

where the dot denotes differentiation. This equation says that $G(\theta)$ takes on nearly the same values as the line L which passes through the point $(t_1, G(t_1))$ with slope $\dot{G}(t_1)$ [i.e., is tangent to the curve $y = G(x)$ at $x = t_1$], provided that θ is not too far from t_1. Solving Equation 1.2 for θ, we see that

$$\theta \approx t_1 - \frac{G(t_1)}{\dot{G}(t_1)},$$

so that a potentially better estimator for θ might be (see Figure 1.1)

$$t_2 = t_1 - \frac{G(t_1)}{\dot{G}(t_1)}. \tag{1.3}$$

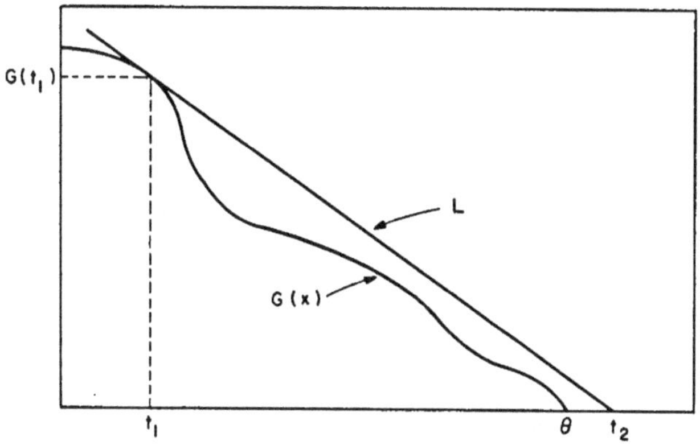

Figure 1.1 Graphical interpretation of Newton's method.

In turn, t_2 could be "improved" in the same way, and Equation 1.3 suggests that an ever-improving sequence of estimators for θ can be obtained by means of the recursion

$$t_{n+1} = t_n - \frac{G(t_n)}{\dot{G}(t_n)} \qquad (n \geq 1). \tag{1.4}$$

It would appear, though, that the first guess t_1 must be close to θ in order that the linear approximation, Equation 1.2, should be accurate. This is not essential if $|\dot{G}|$ is bounded above and away from zero:

$$0 < b \leq |\dot{G}(x)| \leq d < \infty.$$

We choose a number a to satisfy

$$0 < a \leq \frac{b}{d},$$

and we modify the recursion, Equation 1.4, to read

$$t_{n+1} = t_n - a_n G(t_n), \qquad a_n = \frac{a}{\dot{G}(t_n)}. \tag{1.5}$$

It is easy to show that t_n converges to θ as $n \to \infty$. Indeed, by the mean-value theorem, we obtain

$$G(t_n) = \dot{G}(u_n)(t_n - \theta), \tag{1.6}$$

where u_n lies between θ and t_n. Thus, by Equations 1.5 and 1.6, it follows that

$$t_{n+1} - \theta = [1 - a_n\dot{G}(u_n)](t_n - \theta) = \left\{\prod_{j=1}^{n}\left[1 - a\frac{\dot{G}(u_j)}{\dot{G}(t_j)}\right]\right\}(t_1 - \theta). \tag{1.7}$$

But

$$0 \leq 1 - a\frac{d}{b} \leq 1 - a\frac{\dot{G}(u_j)}{\dot{G}(t_j)} \leq 1 - a\frac{b}{d} < 1,$$

so that

$$|t_{n+1} - \theta| \leq \left(1 - a\frac{b}{d}\right)^n |t_1 - \theta| \to 0$$

as $n \to \infty$.

Let us now complicate matters by letting G vary with n. There is a sequence of monotone differentiable functions, G_n, all having a common root θ:

$$G_n(\theta) = 0 \qquad (n = 1, 2, \cdots).$$

Again, we estimate θ by sequences of the form

$$t_{n+1} = t_n - a_n G_n(t_n).$$

In precisely the same way, in place of Equation 1.7 we obtain

$$t_{n+1} - \theta = [1 - a_n \dot{G}_n(u_n)](t_n - \theta) = \left\{\prod_{j=1}^{n} [1 - a_j \dot{G}_j(u_j)]\right\}(t_1 - \theta).$$

Now assuming that

$$0 < b_n < |\dot{G}_n(x)| \leq Mb_n < \infty$$

for all n and all x, we choose a_n so that

1. a_n has the same sign as $\dot{G}_n$,
2. $|a_n| \leq \dfrac{1}{Mb_n}$,
3. $\sum_n |a_n b_n| = \infty$.

Then we have

$$0 \leq \prod_{j=1}^{n} [1 - a_j \dot{G}(u_j)] \leq \prod_{j=1}^{n} (1 - |a_j b_j|) \to 0,$$

and $|t_{n+1} - \theta|$ tends once again to zero as $n \to \infty$.

This technique can be applied to the problem of discrete-time curve fitting: Suppose $Y_1, Y_2, \cdots$ is a sequence of numbers, and it is known that this sequence is one of a family of sequences, $\{F_n(\theta)\}$, indexed by a real parameter θ. Here θ is not known, and we wish to find that value of θ for which

$$Y_n = F_n(\theta) \qquad (n = 1, 2, \cdots).$$

If we let

$$G_n(x) = F_n(x) - Y_n,$$

the desired parameter value is that value of x which makes $G_n(x)$ vanish identically in n.

Now let noise be introduced, so that the sequence of observations, Y_n, are corrupted versions of $F_n(\theta)$:

$$Y_n = F_n(\theta) + W_n \qquad (n = 1, 2, \cdots),$$

where W_n is (zero mean) noise. Motivated by the previous discussion, we consider estimation schemes of the form

$$t_{n+1} = t_n + a_n[Y_n - F_n(t_n)], \tag{1.8}$$

which can be rewritten as

$$t_{n+1} = t_n - |a_n| \, Z_n(t_n). \tag{1.8a}$$

For every x, we can regard $Z_n(x)$ as an observable random variable with expectation equal to

$$G_n(x) = \operatorname{sgn} \dot{F}_n[F_n(x) - F_n(\theta)] = |\dot{F}_n(u_n)|(x - \theta), \tag{1.9}$$

where $u_n = u_n(x, \theta)$ lies between x and θ. Thus,

$$t_{n+1} - \theta = (1 - |a_n \dot{F}_n(u_n)|)(t_n - \theta) + a_n W_n, \tag{1.10}$$

and we are led to the study of certain first-order nonlinear difference equations with stochastic driving terms.

This brings to mind the literature associated with stochastic approximation, which dates back to a paper by Robbins and Monro (1951). That paper concerns itself with the problem of estimating the root, say α, of an unknown (time-homogeneous) regression function $G(x)$, which is the mean value of an observable random variable $Z(x)$. The distribution of the latter depends on a scalar parameter, x, which can be controlled by the experimenter. They proposed that α be estimated recursively by Equation 1.8a, where $Z(t_n)$ is the value of an observation taken at the "level" $x = t_n$, and $\{a_n\}$ is any nonsummable null sequence of scalars with $\sum_n a_n^2 < \infty$. The success of the Robbins–Monro procedure (it converges to α with probability one and in mean square under a wide range of conditions) encourages us to believe in the reasonableness of Equation 1.8.

Burkholder (1956) has studied processes of the form of Equation 1.8a in detail. In fact, he considers the more general situation where the root of G_n depends upon n but converges to a limit θ as $n \to \infty$. (This is not just an academic generalization, for such a result is needed in the treatment of the Kiefer–Wolfowitz procedure for locating the minimum of a time-homogeneous regression function.) Consequently, there will be some overlap between his work and Chapters 2 through 4 of the present work. In fact, after appropriate reinterpretation of the symbols, we obtain some results that are significantly stronger than those given by Burkholder.

If we view the stochastic-approximation literature as a study in the asymptotic behavior of the solutions to a certain class of nonlinear first-order difference equations with stochastic driving terms, then the results of this monograph (particularly Chapters 3 and 4) serve to extend and complement many of the results in that literature, and accounts for our choice of title. However, our primary consideration is nonlinear regression *per se* and, for this reason, we often fail to state theorems with the weakest possible hypotheses; we want to keep their statements and proofs relatively simple.

We will treat the scalar-parameter case, Equation 1.8, and the general vector case, Equation 1.1, separately. For the vector-parameter case, we will treat the topics of strong consistency (probability-one convergence) and mean-square convergence. In the scalar-parameter case, we also treat the questions of convergence rates, asymptotic distribution theory, and efficiency. A wide class of gain sequences are examined. Some are deterministic, and some depend on the actual data which have been observed. Examples are sprinkled throughout the body of the monograph, and Chapter 8 is devoted exclusively to applications.

The techniques we use are, by now, standard to those who are familiar with the literature of stochastic approximation, but for the sake of the nonspecialist we have tried to keep our treatment self-contained. In all cases, we seek the asymptotic properties of the solutions to the intrinsically nonlinear difference equations of the type 1.1. We accomplish this by studying the asymptotic properties of certain linear difference equations which, in a sense, dominate the original ones.

Now a word about notation. In Chapters 6 through 9, we do not adhere to the convention which reserves lower-(resp. upper-)case boldface symbols for vectors (resp. matrices). The reader must keep in mind not only this point but also the orders of the various vectors and matrices involved. The symbol $a_n = O(b_n)$ means that $|a_n/b_n|$ has a finite limit superior as n tends to infinity, while $a_n = o(b_n)$ means the ratio tends to zero. The balance of the abbreviations are standard and are defined when they are first used.

We begin by studying the problems of probability-one and mean-square convergence in the scalar case.

PART I

THE SCALAR-PARAMETER CASE

2. Probability-One and Mean-Square Convergence

2.1 The Basic Assumptions (A1 Through A5″)

Throughout Part I we will use certain assumptions, the first of which is as follows:

A1. $\{Y_n : n = 1, 2, \cdots\}$ is an observable stochastic process of the form $Y_n = F_n(\theta) + W_n$, where $W_1, W_2, \cdots$ have uniformly bounded variances. The function $F_n(\theta)$ is known except for a real parameter θ. However, θ is known to lie in an interval $J = (\xi_1, \xi_2)$, whose end points may be finite or infinite. For each value of n, $F_n(\cdot)$ is assumed to be monotone and differentiable with respect to the argument in parentheses.

If J happens to be finite or semifinite, it is reasonable to constrain estimators for θ so that they always fall in J. To this end, we define the limiting operation

$$[x]_{\xi_1}^{\xi_2} = \begin{cases} \xi_2 & \text{if } x \geq \xi_2, \\ x & \text{if } \xi_1 < x < \xi_2, \\ \xi_1 & \text{if } x \leq \xi_1, \end{cases} \tag{2.1}$$

and, accordingly, will consider truncated estimation recursions of the form

$$t_{n+1} = [t_n + a_n[Y_n - F_n(t_n)]]_{\xi_1}^{\xi_2} \qquad (n = 1, 2, \cdots; t_1 \text{ arbitrary}). \tag{2.2}$$

In the work that follows, we will use certain symbols and assumptions (in addition to A1) repeatedly. For the sake of future brevity and ease of comparison, we list them here once and for all and will refer to them later by number.

A1′. In addition to Assumption A1, $W_1, W_2, \cdots$ is a zero-mean independent process.

A2. For every $n \geq 1$, $d_n = \sup_{x \in J} |\dot{F}_n(x)| < \infty$, where $\dot{F}_n$ denotes the derivative of F_n.

A3. $B_n^2 = \sum_{k=1}^{n} b_k^2 \to \infty$ with n, where $b_k = \inf_{x \in J} |\dot{F}_k(x)|$.

A4. $\sup_n \frac{d_n}{b_n} < \infty$.

A5. $\limsup_n \frac{d_n^2}{B_n^2} < 1$.

A5′. $\limsup_n \left(\frac{d_n}{b_n}\right)\left(\frac{d_n^2}{B_n^2}\right) < 1$.

A5″. $\lim_n \frac{b_n^2}{B_n^2} = 0$.

A5‴. $\sum_n \left(\frac{b_n^2}{B_n^2}\right)^2 < \infty$.

We note that Assumption A5‴ implies A5″, Assumptions A4 and A5″ imply A5′, and Assumption A5′ implies A5.

These assumptions are stated in terms of the quantities b_n and d_n, which are, in turn, defined in terms of an interval J that is known to include the true value of the parameter. Obviously, J should be chosen to be the smallest interval known to contain θ. (In general, the smaller J is, the weaker are Assumptions A2 through A5‴.) In the absence of prior knowledge, J can be (must be) taken to be the real line.

Even when J is a finite interval, it is not essential that the estimator sequence be truncated. We could alternatively redefine the regression function outside of J by linearity and then use an untruncated procedure to estimate θ. That is to say, we could define

$$F_n^*(x) = \begin{cases} F_n(\xi_2) + \dot{F}_n(\xi_2)(x - \xi_2) & \text{if } x \geq \xi_2, \\ F_n(x) & \text{if } \xi_1 < x < \xi_2, \\ F_n(\xi_1) + \dot{F}_n(\xi_1)(x - \xi_1) & \text{if } x \leq \xi_1, \end{cases}$$

and we could use the untruncated scheme

$$t_{n+1}^* = t_n^* + a_n[Y_n - F_n^*(t_n)].$$

Since we know that

$$\sup_{\xi_1 < x < \xi_2} |\dot{F}_n(x)| = \sup_{-\infty < x < \infty} |\dot{F}_n^*(x)|$$

and

$$\inf_{\xi_1 < x < \xi_2} |\dot{F}_n(x)| = \inf_{-\infty < x < \infty} |\dot{F}_n^*(x)|,$$

any of the Assumptions A2 through A5‴ that hold for $F_n(\cdot)$ over $J = (\xi_1, \xi_2)$ will also hold for $F_n^*(\cdot)$ over $J^* = (-\infty, \infty)$. Hence, the results of this chapter (as well as the next) will apply to the untruncated estimators t_n^* whenever they apply to the truncated ones, t_n. In most applications, however, common sense seems to dictate that we should use truncated procedures whenever we can.

The first theorem demonstrates the strong consistency of the estimation sequence, Equation 2.2, for a wide class of gain sequences. [For $J = (-\infty, \infty)$, independent observations and gains which do not depend on the iterates, the result becomes Burkholder's (1956) Theorem 1 after an appropriate interpretation of the symbols.]

2.2 Theorems Concerning Probability-One and Mean-Square Convergence for General Gains

THEOREM 2.1

Let $\{Y_n : n = 1, 2, \cdots\}$ be an observable process satisfying Assumptions A1 and A2. Let $\{t_n\}$ be a sequence of estimators defined by the recursion

$$t_{n+1} = [t_n + a_n(t_1, t_2, \cdots, t_n)[Y_n - F_n(t_n)]]_{\xi_1}^{\xi_2} \qquad (t_1 \text{ arbitrary}),$$

where, for each n, $a_n(\cdot)$ is a Borel function of n real variables, and $J = (\xi_1, \xi_2)$ is any interval, finite or infinite, known to include the true parameter value θ. Let $J^{(n)}$ be the n-dimensional cube whose sides are the interval J. If the gain sequence $\{a_n(\cdot)\}$ is chosen so that

1. For each n, the sign of $a_n(\mathbf{x})$ is constant over $J^{(n)}$ and equal to the sign of $\dot{F}_n(\cdot)$,
2. $\sup_{\mathbf{x} \in J^{(n)}} |a_n(\mathbf{x})| < \dfrac{1}{d_n}$ for all suitably large values of n,

and

3. $\sum_n b_n \inf_{\mathbf{x} \in J^{(n)}} |a_n(\mathbf{x})| = \infty$,

then t_n converges to θ with probability one if either

4. $\sum_n \sup_{\mathbf{x} \in J^{(n)}} |a_n(\mathbf{x})| < \infty$

or

5. $\sum_n \sup_{\mathbf{x} \in J^{(n)}} |a_n(\mathbf{x})|^2 < \infty$ and Assumption A1′ holds.

Proof. For notational convenience, denote

$$a_n(t_1, t_2, \cdots, t_n) \quad \text{by} \quad a_n,$$

$$\inf_{\mathbf{x} \in J^{(n)}} |a_n(\mathbf{x})| \quad \text{by} \quad \inf |a_n|,$$

$$\sup_{\mathbf{x} \in J^{(n)}} |a_n(\mathbf{x})| \quad \text{by} \quad \sup |a_n|.$$

Let

$$T_n = T_n(t_1, t_2, \cdots, t_n) = t_n + a_n[F_n(\theta) - F_n(t_n)] \tag{2.3}$$

and

$$Z_n = a_n[Y_n - F_n(\theta)]. \tag{2.4}$$

Then we obtain

$$|t_{n+1} - \theta| = |[T_n + Z_n]_{\xi_1}^{\xi_2} - \theta|,$$

and, consequently,

$$|t_{n+1} - \theta| \le |T_n - \theta + Z_n|. \tag{2.5}$$

Indeed, Equation 2.5 clearly holds if $(T_n + Z_n) \in J$. Otherwise, because $\xi_1 < \theta < \xi_2$, θ is closer to the end point of J nearest to $T_n + Z_n$ than it is to $T_n + Z_n$.

The placement of the absolute-value signs on the right-hand side makes it awkward to iterate the Inequality 2.5. However, suppose that we can choose a positive null sequence of real numbers $\{A_n\}$ such that

$$\lim_n \frac{Z_n}{A_n} = 0 \qquad \text{a.s.}, \tag{2.6}$$

that is, with probability one. Then, by Condition 2, we can choose N so large that

$$d_n \sup |a_n| < 1 \tag{2.7}$$

and

$$|Z_n| \le \frac{A_n}{2} \tag{2.8}$$

both hold whenever $n \ge N$ (N a random variable). Fix $n \ge N$. If $|T_n - \theta| < A_n/2$, then by Equation 2.5 it follows that

$$|t_{n+1} - \theta| \le A_n.$$

In the contrary case,

$$|T_n - \theta| > \frac{A_n}{2} \geq |Z_n|,$$

which implies that the sign of $(T_n - \theta + Z_n)$ is equal to the sign of $(T_n - \theta)$:

$$\begin{aligned} |T_n - \theta + Z_n| &= (T_n - \theta + Z_n) \operatorname{sgn} (T_n - \theta + Z_n) \\ &= (T_n - \theta + Z_n) \operatorname{sgn} (T_n - \theta) \\ &= |T_n - \theta| + Z_n \operatorname{sgn} (T_n - \theta). \end{aligned}$$

Setting

$$X_n = Z_n \operatorname{sgn} (T_n - \theta), \tag{2.9}$$

we have, in this case,

$$|t_{n+1} - \theta| \leq |T_n - \theta| + X_n.$$

In either event, therefore,

$$|t_{n+1} - \theta| \leq \max \{A_n, |T_n - \theta| + X_n\} \tag{2.10}$$

if $n \geq N$, and this is the key relationship for our subsequent analysis.

To establish Equation 2.6, we choose a positive null sequence $\{A_n\}$ so that

$$\sum_n \frac{(\sup |a_n|)^2}{A_n^2} < \infty.$$

This is always possible since

$$\sum_n (\sup |a_n|)^2 < \infty$$

under either Condition 4 or 5. But then, from Equation 2.4, we obtain

$$\sum_n \mathscr{E} \left(\frac{Z_n}{A_n}\right)^2 \leq \text{const} \sum_n \left(\frac{\sup |a_n|}{A_n}\right)^2 < \infty,$$

so that

$$\mathscr{E} \sum_n \left(\frac{Z_n}{A_n}\right)^2 < \infty.$$

It follows from the monotone convergence theorem (Loève, 1960, p. 152) that

$$\sum_n \left(\frac{Z_n}{A_n}\right)^2 < \infty \qquad \text{a.s.},$$

which in turn implies Equation 2.6.

Returning our attention to Equation 2.10, we notice that

$$|T_n - \theta| = |t_n - \theta + a_n[F_n(\theta) - F_n(t_n)]|$$

by virtue of Equation 2.3. Using the mean-value theorem, we find that

$$|T_n - \theta| = |[1 - a_n\dot{F}_n(u_n)](t_n - \theta)|,$$

where u_n lies between θ and t_n. By Condition 1, we have $a_n\dot{F}_n > 0$. Thus, in view of Equation 2.7, it follows that

$$0 < b_n \inf |a_n| \leq a_n\dot{F}_n(u_n) \leq d_n \sup |a_n| < 1$$

if $n \geq N$, so that

$$|T_n - \theta| \leq (1 - b_n \inf |a_n|)|t_n - \theta|.$$

This combines with Equation 2.10 to give, for all such indices,

$$|t_{n+1} - \theta| \leq \max \{A_n, (1 - b_n \inf |a_n|)|t_n - \theta| + X_n\}. \quad (2.11)$$

If Equation 2.11 is iterated from n back to N, we obtain

$$|t_{n+1} - \theta| \leq \max \left[\max_{N \leq i \leq n} \left(\frac{A_i P_n}{P_i} + \sum_{k=i+1}^{n} \frac{P_n X_k}{P_k} \right), \right.$$
$$\left. \frac{P_n}{P_{N-1}} |t_N - \theta| + \sum_{k=N}^{n} \frac{P_n X_k}{P_k} \right] \quad (2.12)$$

(which can be verified by induction), where

$$P_n = \prod_{j=1}^{n} (1 - b_j \inf |a_j|).$$

Since $1 - x \leq e^{-x}$ (for all x), we see that

$$\frac{P_n}{P_{N-1}} \leq \exp [-\sum_{j=N}^{n} b_j \inf |a_j|] \to 0$$

as $n \to \infty$ by Condition 3.

We still have to show that

$$\sum_k X_k < \infty \qquad \text{a.s.}, \quad (2.13)$$

for then, by Lemma 2 of the Appendix, it follows that

$$\max_{N \leq i \leq n} \sum_{k=i+1}^{n} \frac{P_n}{P_k} X_k \to 0 \qquad \text{a.s. as} \quad n \to \infty.$$

Since

$$\max_{N \leq i \leq n} \left| \frac{A_i P_n}{P_i} \right| \leq \max_{N \leq i \leq n} |A_i| \to 0$$

and

$$\frac{P_n}{P_{N-1}} |t_N - \theta| \to 0 \qquad \text{as} \quad n \to \infty,$$

the desired conclusion will follow from Equation 2.12.

To establish Equation 2.13, we can use either Condition 4 or 5. Under Condition 4,

$$\sum_k \mathscr{E}|X_k| \le \text{const} \sum_k [\sup |a_k|] < \infty.$$

Then $\mathscr{E} \sum_k |X_k| < \infty$ and, hence, $\sum_k X_k < \infty$ a.s. by the monotone-convergence theorem. Under Condition 5, we notice that the random variables $X_1, X_2, \cdots, X_k$ are functions of $Z_1, \cdots, Z_k, T_1, \cdots, T_k$, which are themselves functions of $t_1, t_2, \cdots, t_k, W_1, \cdots, W_k$, where $Z_k = a_k W_k$ (see Equations 2.3, 2.4, and 2.9). In turn, $t_1, \cdots, t_k$ are functions of $W_1, \cdots, W_{k-1}$, so that the Borel field induced by $X_1, \cdots, X_k$ is a subfield of the Borel field induced by $W_1, \cdots, W_k$. Thus,

$$\mathscr{E}(X_{k+1}|X_k, \cdots, X_1) = \mathscr{E}[\mathscr{E}(X_{k+1}|W_k, \cdots, W_1)|X_k, \cdots, X_1].$$

The inner expectation is equal to

$$a_{k+1}(t_1, \cdots, t_{k+1}) \operatorname{sgn}(T_{k+1} - \theta)\mathscr{E}(W_{k+1}|W_k, \cdots, W_1) = 0$$

by virtue of the assumed independence of the W's. Thus,

$$\mathscr{E}(X_{k+1}|X_k, \cdots, X_1) = 0 \qquad \text{a.s.}$$

and, since

$$\mathscr{E}X_{k+1}^2 \le \text{const} \sup |a_{k+1}|^2,$$

we see that

$$\sum_k \mathscr{E}X_k^2 < \infty.$$

Theorem D on page 387 of Loève (1960) applies, thereby proving Equation 2.13. Q.E.D.

The conditions for mean-square convergence are identical with those required for probability-one convergence, although the method of proof differs.

THEOREM 2.2

Let $\{Y_n : n = 1, 2, \cdots\}$ be an observable process satisfying Assumptions A1 and A2. Let $\{t_n\}$ be a sequence of estimators defined by the recursion

$$t_{n+1} = [t_n + a_n(t_1, t_2, \cdots, t_n)[Y_n - F_n(t_n)]]_{\xi_1}^{\xi_2} \qquad (t_1 \text{ arbitrary}),$$

where $a_n(\cdot)$ is a Borel function of n real variables and $J = (\xi_1, \xi_2)$ is any interval, finite or infinite, known to include the true parameter value θ. If Conditions 1, 2, 3, and either 4 or 5 of Theorem 2.1 hold, then

$$\lim_{n\to\infty} \mathscr{E}(t_n - \theta)^2 = 0.$$

Proof. By Equation 2.5, we see that

$$(t_{n+1} - \theta)^2 \le (T_n - \theta)^2 + 2Z_n(T_n - \theta) + Z_n^2, \tag{2.14}$$

where, as previously,

$$T_n = t_n + a_n[F_n(\theta) - F_n(t_n)], \qquad Z_n = a_n[Y_n - F_n(\theta)] = a_n W_n.$$

Furthermore, by virtue of Assumption A1,

$$(\mathscr{E} W_n)^2 \le \mathscr{E} W_n^2 \le \sigma^2 < \infty \qquad \text{for all } n. \tag{2.15}$$

By the mean-value theorem,

$$T_n - \theta = [1 - a_n \dot{F}_n(u_n)](t_n - \theta),$$

so that for large n,

$$(T_n - \theta)^2 \le (1 - b_n \inf |a_n|)^2 (t_n - \theta)^2. \tag{2.16}$$

Suppose that Condition 4 holds. Then, combining Equations 2.14, 2.15, and 2.16, and letting

$$e_n^2 = \mathscr{E}(t_n - \theta)^2, \tag{2.17}$$

we obtain

$$\begin{aligned} 0 \le e_{n+1}^2 &\le (1 - b_n \inf |a_n|)^2 e_n^2 + \sigma^2 \sup |a_n|^2 + 2\sigma e_n \sup |a_n| \\ &\le (1 - b_n \inf |a_n|)^2 e_n^2 + M_1 \sup |a_n|(1 + e_n), \end{aligned} \tag{2.18}$$

where M's will denote various constants. Since $\sup |a_n|$ is summable and since $0 < b_n \inf |a_n| < 1$ for large n by Condition 2, Lemma 3 of the Appendix can be applied to give

$$\sup_n e_n^2 = M_2^2 < \infty.$$

Thus, from Equation 2.18, we obtain

$$0 \le e_{n+1}^2 \le (1 - b_n \inf |a_n|)^2 e_n^2 + M_3 \sup |a_n|. \tag{2.19}$$

Choose N so large that

$$b_n \inf |a_n| < 1 \qquad \text{for} \quad n \ge N,$$

and iterate Equation 2.19 back to N. We get

$$0 \le e_{n+1}^2 \le \frac{P_n^2}{P_{N-1}^2} e_N^2 + \sum_{k=N}^{n} \frac{P_n^2}{P_k^2} M_3 \sup |a_k|, \tag{2.20}$$

where

$$P_n = \prod_{j=1}^{n} (1 - b_j \inf |a_j|) \to 0.$$

A special case of Lemma 2 in the Appendix is (a version of the Kronecker Lemma):

$$\lim_{n\to\infty} \sum_{k=N}^{n} \left(\frac{P_n}{P_k}\right) \sup |a_k| = 0, \tag{2.21}$$

since $\sup |a_k|$ is summable. But since $0 < P_n/P_k < 1$ for all $N \leq k \leq n$, it follows that $P_n^2/P_k^2 < P_n/P_k$. This and Equation 2.21, together with Equation 2.20, give $e_n^2 \to 0$.

Under Condition 5, the W_n are independent, so that for every n, W_n is independent of $a_n(t_1, \cdots, t_n)$ and $T_n(t_1, \cdots, t_n)$. Thus,

$$\begin{aligned} \mathscr{E} Z_n(T_n - \theta) &= \mathscr{E}[\mathscr{E}(Z_n(T_n - \theta) | W_1, \cdots, W_{n-1})] \\ &= \mathscr{E}[a_n(T_n - \theta)\mathscr{E}(W_n | W_1, \cdots, W_{n-1})] = 0. \end{aligned} \tag{2.22}$$

By Equations 2.14, 2.16, and 2.22,

$$\mathscr{E}(t_{n+1} - \theta)^2 \leq (1 - b_n \inf |a_n|)^2 \mathscr{E}(t_n - \theta)^2 + \mathscr{E} Z_n^2, \tag{2.23}$$

and, by Equation 2.15,

$$0 \leq e_{n+1}^2 \leq (1 - b_n \inf |a_n|)^2 e_n^2 + (\sigma \sup |a_n|)^2. \tag{2.24}$$

After iterating back to N, we have

$$0 \leq e_{n+1}^2 \leq \left(\frac{P_n}{P_{N-1}}\right)^2 e_N^2 + \sigma^2 \sum_{k=N}^{n} \left(\frac{P_n}{P_k}\right)^2 (\sup |a_k|)^2. \tag{2.25}$$

Since $P_n \to 0$, the first term tends to zero, and the same argument used earlier shows that

$$\lim_{n\to\infty} \sum_{k=N}^{n} \left(\frac{P_n}{P_k}\right)^2 (\sup |a_k|)^2 = 0.$$

Thus, under either Condition 4 or 5, we have $e_n^2 \to 0$ as $n \to \infty$. Q.E.D.

The conditions of Theorems 2.1 and 2.2 are satisfied by a number of gain sequences, provided that the regression function satisfies a certain number of the assumptions A2 through A5‴ listed at the beginning of this chapter.

2.3 The Prototype Deterministic Gain

Consider the gain sequence

$$a_n(x_1, x_2, \cdots, x_n) = \frac{b_n}{B_n^2} \operatorname{sgn} \dot{F}_n. \tag{2.26}$$

Since $F_n(\cdot)$ is monotone for each n, the sign of $\dot{F}_n(x)$ is independent of x and Equation 2.26 does not depend on the arguments. In instances where speed of real-time computation is an important factor, these deterministic gains possess the virtue of being computable in advance of the data acquisition (although there is the possibility of a storage problem). Since

$$\sup_{\mathbf{x} \in J^{(n)}} a_n(\mathbf{x}) = \inf_{\mathbf{x} \in J^{(n)}} a_n(\mathbf{x}) = a_n,$$

Condition 2 of Theorems 2.1 and 2.2 holds under Assumption A5. Furthermore, Condition 3 is ensured by Assumption A3, because the Abel–Dini Theorem (Knopp, 1947, p. 290) says that

$$\sum_{k=1}^{\infty} \frac{b_k^2}{B_k^{2+r}} \text{ is } \begin{cases} \text{divergent when } r \leq 0 \\ \text{convergent when } r > 0 \end{cases} \quad \text{if} \quad B_n^2 \to \infty. \tag{2.27}$$

(This theorem will be used repeatedly.) If $\{Y_n\}$ is an independent process, Condition 5 also holds under Assumption A3, because

$$\sum_n (\sup a_n)^2 = \sum_n \frac{b_n^2}{B_n^4} < \infty.$$

If $\{Y_n\}$ is not an independent process, Condition 4 holds when

$$b_n \sim n^{\alpha} \qquad (\alpha > 0)$$

(by which we mean that the ratio of the two sides has a nonzero limit inferior and finite limit superior), for then

$$B_n^2 \sim n^{1+2\alpha} \qquad \text{and} \qquad |a_n| = \frac{b_n}{B_n^2} \sim \frac{1}{n^{1+\alpha}},$$

which is summable. In particular, we can do nonlinear polynomial regression,

$$F_n(\theta) = \sum_{j=0}^{p} f_j(\theta) n^j \qquad (p \geq 1),$$

when the errors are dependent and have nonzero means.

2.4 Reduction in the Linear Case

It is instructive to examine our iterative scheme when the regression function is linear:

$$F_n(\theta) = b_n \theta \qquad \text{and} \qquad a_n = \frac{b_n}{B_n^2}.$$

(Note the slightly different usage of b_n here.) We take $J = (-\infty, \infty)$, and the recursion Equation 2.2 becomes

$$\begin{aligned} t_{n+1} &= t_n + a_n[Y_n - b_n t_n] \\ &= (1 - a_n b_n)t_n + a_n Y_n \\ &= \frac{B_{n-1}^2}{B_n^2} t_n + \frac{b_n}{B_n^2} Y_n. \end{aligned}$$

Iterating back to $n = 1$, we get

$$t_{n+1} = \left[\prod_{j=1}^{n} \left(\frac{B_{j-1}^2}{B_j^2}\right)\right] t_1 + \sum_{k=1}^{n} \left(\prod_{j=k+1}^{n} \frac{B_{j-1}^2}{B_j^2}\right) \frac{b_k}{B_k^2} Y_k.$$

If the initial (no data) estimate is $t_1 = 0$, then

$$t_{n+1} = \left(\sum_{k=1}^{n} b_k Y_k\right) \Big/ \left(\sum_{k=1}^{n} b_k^2\right),$$

which is precisely the least-squares estimator for θ based upon the first n observations. In other words, the gain sequence, Equation 2.26, yields a recursively defined estimator sequence which is identical to the corresponding sequence of least-squares estimators.

The variance of the least-squares estimate in the case of independent identically distributed residuals is easily computed. Since

$$Y_k = b_k\theta + W_k$$

with $\mathscr{E} W_k^2 = \sigma^2$, it follows that

$$\mathscr{E}(t_{n+1} - \theta)^2 = \frac{\sigma^2}{B_n^2}.$$

Thus, t_n converges in quadratic mean to θ if and only if $B_n^2 \to \infty$. But we have already shown in the preceding paragraph that this condition implies Conditions 3 and 5. Since in the present case $\sup_x |\dot{F}_n(x)| = b_n$, Conditions 1 and 2 also hold. In short, Conditions 1, 2, 3, and 5 of Theorem 2.2 are necessary as well as sufficient conditions for the quadratic-mean (and, we might add, almost sure) convergence of the recursively defined least-squares estimator,

$$t_{n+1} = t_n + \frac{b_n}{B_n^2}(Y_n - b_n t_n) \qquad (t_1 = 0),$$

in the case of a linear regression function.

2.5 Gains That Use Prior Knowledge

In practical applications, prior knowledge about the true value of θ is often available. There may, for example, be a nominal value θ_0, predicted by rough theoretical calculations, which will hopefully be close to the true value of θ. Under this supposition, the regression function can be approximated by the first few terms of its Taylor series expansion:

$$F_n(\theta) \approx F_n(\theta_0) + \dot{F}_n(\theta_0)(\theta - \theta_0).$$

If W_n denotes the error in the nth observation, we can write

$$Y_n \approx F_n(\theta_0) + \dot{F}_n(\theta_0)(\theta - \theta_0) + W_n,$$

or equivalently,

$$Y_n^* \approx \dot{F}_n(\theta_0)\theta + W_n,$$

where

$$Y_n^* = Y_n - [F_n(\theta_0) - \theta_0 \dot{F}_n(\theta_0)].$$

The parameter θ now occurs (approximately) linearly in the mean value of the observable Y_n^*, so that the recursive version of the linear least-squares estimator discussed earlier seems appropriate. Accordingly, the gain sequence would be

$$a_n = \frac{\dot{F}_n(\theta_0)}{\sum_{k=1}^{n} \dot{F}_k^2(\theta_0)}, \tag{2.28}$$

and the iteration would be

$$t_{n+1} = t_n + a_n[Y_n^* - \dot{F}_n(\theta_0)t_n].$$

But (we are, of course, being heuristic)

$$Y_n^* - \dot{F}_n(\theta_0)t_n \approx Y_n - F_n(t_n),$$

so we are led to substitute $Y_n - F_n(t_n)$ for $Y_n^* - \dot{F}_n(\theta_0)t_n$, and to consider the "improved" recursion

$$t_{n+1} = t_n + a_n[Y_n - F_n(t_n)] \qquad (t_1 = \theta_0)$$

(or, perhaps, a truncated version). This particular recursive estimation scheme, that is, the gain sequence, Equation 2.28, is widely used in current practice, especially the vector version (see Section 7.5 in Chapter 7).

2.6 Random Gains

From the theoretical point of view, the gain sequences, Equations 2.26 and 2.28, are deterministic special cases of those of the form

$$a_n(x_1, x_2, \cdots, x_n) = \frac{\alpha_n^2(x_1, x_2, \cdots, x_n) \operatorname{sgn} \dot{F}_n}{\beta_n(x_1, \cdots, x_n) \sum_{k=1}^{n} \gamma_k^2(x_1, \cdots, x_k)}, \tag{2.29}$$

with

$$b_n \leq \alpha_n(\mathbf{x}), \beta_n(\mathbf{x}), \gamma_n(\mathbf{x}) \leq d_n$$

for all $\mathbf{x} \in J^{(n)}$. Some cases in point are the following:

$$a_n(x_1, \cdots, x_n) = \begin{cases} \dfrac{\dot{F}_n(x_n)}{B_n^2} & (\alpha_n = \beta_n = |\dot{F}_n(x_n)|, \quad \gamma_k = b_k), \\ \dfrac{b_n^2}{\dot{F}_n(x_n)B_n^2} & (\alpha_n = b_n, \beta_n = |\dot{F}_n(x_n)|, \quad \gamma_k = b_k), \end{cases}$$

and, more important,

$$a_n(x_1, \cdots, x_n) = \frac{\dot{F}_n(x_n)}{\sum_{k=1}^{n} \dot{F}_k^2(x_k)}. \tag{2.30}$$

The last is referred to as the "adaptive" gain and is quite often used in practice.

The convergence properties of estimates based on gains of the type of Equation 2.29 are determined by considerations of the following sort. If Assumptions A3, A4, and A5′ hold, then Conditions 2, 3, and 5 of Theorems 2.1 and 2.2 hold. Indeed, we see that

$$d_n \sup |a_n| \leq \frac{d_n^3}{b_n B_n^2} < 1$$

and

$$b_n \inf |a_n| \geq \frac{b_n^3}{d_n \sum_{k=1}^{n} d_k^2} \geq \frac{K b_n^2}{B_n^2}.$$

The Abel–Dini theorem, Equation 2.27, therefore guarantees the non-summability of $b_n \inf |a_n|$, as well as the summability of

$$(\sup |a_n|)^2 \leq \left(\frac{d_n^2}{b_n B_n^2}\right)^2 \leq K' \frac{b_n^2}{B_n^4}.$$

Assumption A5′ is not essential if we modify the gain sequence slightly. Let $\{\mu_n\}$ be a positive null sequence, chosen so that

$$\sum_n \frac{\mu_n b_n^2}{B_n^2} = \infty.$$

(This is always possible since $\sum_n b_n^2/B_n^2 = \infty$ under Assumption A3.) Then, set

$$a_n^*(x_1, \cdots, x_n) = \mu_n a_n(x_1, \cdots, x_n).$$

Under Assumptions A3 and A4, it is easy to show that

$$d_n \sup |a_n^*| \le \frac{K\mu_n b_n^2}{B_n^2} \le K\mu_n \to 0,$$

$$\sum_n b_n \inf |a_n^*| \ge K' \sum_n \frac{\mu_n b_n^2}{B_n^2} = \infty,$$

and

$$\sum_n [\sup |a_n^*|]^2 = \sum_n \mu_n^2 [\sup |a_n|]^2 \le K \sum_n \frac{\mu_n^2 b_n^2}{B_n^4} < \infty.$$

Thus, the modified gain sequence

$$a_n^*(x_1, \cdots, x_n) = \mu_n a_n(x_1, \cdots, x_n)$$

satisfies Conditions 1, 2, 3, and 5 of Theorems 2.1 and 2.2 if a_n is of the form of Equation 2.29. Hence, the sequence

$$t_{n+1} = [t_n + a_n^*[Y_n - F_n(t_n)]]_{\xi_1}^{\xi_2}$$

will be a consistent estimator sequence if $\{Y_n\}$ is an independent process whose regression function satisfies Assumptions A2, A3, and A4.

A still broader class of gain sequences satisfy the conditions

$$a \frac{b_n}{B_n^2} \le |a_n(\mathbf{x})| \le a' \frac{b_n}{B_n^2} \qquad \text{for all } \mathbf{x} \in J^{(n)}, \qquad \operatorname{sgn} a_n = \operatorname{sgn} \dot{F}_n.$$

If we impose Assumptions A3, A4, and A5″, then it is easy to show that

$$d_n \sup |a_n| \to 0.$$

The same arguments used in the previous paragraph apply to the respective nonsummability and summability of

$$\sum_n b_n \inf |a_n| \qquad \text{and} \qquad \sum_n (\sup |a_n|)^2.$$

Again, Assumption A5″ can be dispensed with if a_n is replaced by $a_n^* = \mu_n a_n$, where μ_n is a positive null sequence chosen so that

$$\sum_n \frac{\mu_n b_n^2}{B_n^2} = \infty.$$

We summarize all the foregoing (in what is actually a corollary to Theorems 2.1 and 2.2) in two more theorems.

2.7 Theorems Concerning Probability-One and Mean-Square Convergence for Particular Gains; Application to Polynomial Regression

THEOREM 2.3

Let $\{Y_n:n = 1, 2, \cdots\}$ be a stochastic process satisfying Assumptions A1′, A2, and A3. Let

$$t_{n+1} = [t_n + a_n(t_1, \cdots, t_n)[Y_n - F_n(t_n)]]_{\xi_1}^{\xi_2},$$

where $J = (\xi_1, \xi_2)$ is an interval containing the true parameter θ. Then, as $n \to \infty$, $t_n \to \theta$ with probability one and in mean square if

1. $a_n = \operatorname{sgn} \dot{F}_n \dfrac{b_n}{B_n^2}$, and Assumption A5 holds,

or

2. $$a_n(x_1, \cdots, x_n) = \frac{\alpha_n^2(x_1, \cdots, x_n) \operatorname{sgn} \dot{F}_n}{\beta_n(x_1, \cdots, x_n) \sum_{j=1}^{n} \gamma_j^2(x_1, \cdots, x_j)},$$

   where $b_n \le \alpha_n(\mathbf{x}), \beta_n(\mathbf{x}), \gamma_n(\mathbf{x}) \le d_n$ for all $\mathbf{x} \in J^{(n)}$, and Assumptions A4 and A5′ hold,

or

3. $\dfrac{ab_n}{B_n^2} \le |a_n(\mathbf{x})| \le \dfrac{a'b_n}{B_n^2}$ for some $0 < a \le a' < \infty$, $\operatorname{sgn} a_n(\mathbf{x}) = \operatorname{sgn} \dot{F}_n$ for all $\mathbf{x} \in J^{(n)}$, and Assumptions A4 and A5″ hold.

Furthermore, if $a_n(x_1, \cdots, x_n)$ is replaced by $a_n^*(x_1, \cdots, x_n) = \mu_n a_n(x_1, \cdots, x_n)$, where μ_n is a positive null sequence chosen so that

$$\sum_n \frac{\mu_n b_n^2}{B_n^2} = \infty,$$

then Assumptions A5, A5′, and A5″ can be dispensed with in Conditions 1, 2, and 3, respectively.

For the special case of polynomial regression, most of the conditions are automatically satisfied and the independence assumption can be dropped.

THEOREM 2.4

Let $\{Y_n:n = 1, 2, \cdots\}$ be a stochastic process satisfying Assumption A1, where

$$F_n(\theta) = \sum_{j=0}^{p} f_j(\theta) n^j \qquad (\theta \in J, \mathrm{p} \ge 1),$$

$f_p(\cdot)$ is monotone on J, and

$$0 < \inf_{x \in J} |\dot{f}_j(x)| \le \sup_{x \in J} |\dot{f}_j(x)| < \infty \qquad (j = 0, 1, \cdots, p).$$

If, for some $0 < a \le a' < \infty$ and all $\mathbf{x} \in J^{(n)}$,

$$\frac{ab_n}{B_n^2} \le |a_n(\mathbf{x})| \le \frac{a'b_n}{B_n^2}$$

and

$$\operatorname{sgn} a_n(\mathbf{x}) = \operatorname{sgn} \dot{f}_p,$$

then the estimator

$$t_{n+1} = [t_n + a_n(t_1, \cdots, t_n)[Y_n - F_n(t_n)]]_{\xi_2}^{\xi_1} \qquad (t_1 \text{ arbitrary})$$

converges to θ with probability one and in mean square as $n \to \infty$.

Proof. Denote

$$\sup_{x \in J} |\dot{f}_j(x)| \qquad \text{by} \quad \sup |\dot{f}_j|,$$

$$\inf_{x \in J} |\dot{f}_j(x)| \qquad \text{by} \quad \inf |\dot{f}_j|.$$

If f_p is nondecreasing,

$$\inf_{x \in J} \dot{F}_n(x) \ge n^p \inf |\dot{f}_p|[1 + o(1)]$$

as $n \to \infty$, and if f_p is nonincreasing, we have

$$\sup_{x \in J} \dot{F}_n(x) \le -n^p \sup |\dot{f}_p|[1 + o(1)].$$

In either case, $F_n(\cdot)$ is monotone for large n, and it is easy to find constants $0 < K_1 \le K_2 < \infty$ and N such that

$$K_1 n^p \le \inf_{x \in J} |\dot{F}_n(x)| \le \sup_{x \in J} |\dot{F}_n(x)| \le K_2 n^p$$

whenever $n \ge N$. Thus, Assumption A1 and Condition 1 of Theorems 2.1 and 2.2 hold when $n \ge N$. Conditions 2, 3, and 4 hold automatically. Q.E.D.

Naturally, there are regressions (some of great practical importance) which fail to satisfy the conditions we require in order to perform recursive estimation procedures. We will exhibit two cases where one or more of the conditions of Theorems 2.1 and 2.2 are violated.

2.8 Trigonometric Regression

For

$$F_n(\theta) = \cos n\theta \qquad (0 < \theta < \pi),$$

the monotonicity restriction is violated: $\dot{F}_n(\theta) = -n \sin n\theta$ changes sign at least once for every $n \geq 2$ as θ varies over $J = (0, \pi)$. Fortunately, other computationally convenient estimators are available. For example, we can estimate $\cos \theta$ in a consistent (both mean-square and probability-one) fashion, using the estimator

$$t_n = \frac{C_{2n} + \sqrt{C_{2n}^2 + 8C_{1n}^2}}{4C_{1n}},$$

where C_{1n} and C_{2n} are the sample autocovariance functions at lags one and two, respectively. These can, of course, be computed recursively in n. Knowledge of $\cos \theta$ is tantamount to knowledge of θ when $0 < \theta < \pi$. [When J is $(0, 2\pi)$, an independent estimate of $\sin \theta$ is needed to resolve the ambiguity in the angle.] This problem and various more realistic generalizations of it (e.g., unknown phase and amplitude) are the subject of a planned paper by the second-named author.

2.9 Exponential Regression

The function

$$F_n(\theta) = e^{n\theta} \qquad (0 < \xi_1 < \theta < \xi_2 < \infty)$$

violates Conditions 2 and 3 of Theorems 2.1 and 2.2 in an essential way. For, if $a_n(\mathbf{x})$ is any gain satisfying Condition 2, it follows that

$$b_n \inf_{\mathbf{x} \in J^{(n)}} |a_n(\mathbf{x})| \leq \exp[n\xi_1] \exp[-n\xi_2] = \exp[-n(\xi_2 - \xi_1)],$$

and this is always summable.

However, common sense tells us that if the noise variance is bounded, one should be able to estimate θ with ease and accuracy (plot the data on semilog paper), because the "signal-to-noise ratio" grows exponentially fast. This is indeed the case. Suppose

$$Y_n = e^{n\theta} + W_n \qquad \text{and} \qquad \sup_n \mathscr{E} W_n^2 < \infty.$$

We let

$$Y_n^* = \begin{cases} \log Y_n & \text{if } Y_n > A, \\ \log A & \text{if } Y_n \leq A, \end{cases}$$

where A is a positive constant chosen to save us the embarrassment of having to take logarithms of numbers near and less than zero. Then

$$Y_n^* = n\theta + V_n,$$

where

$$V_n = \begin{cases} \log(1 + e^{-n\theta} W_n) & \text{if } Y_n > A, \\ \log Ae^{-n\theta} & \text{if } Y_n \leq A. \end{cases}$$

With high probability, Y_n is going to be larger than A when n is large, so that (heuristically)

$$Y_n^* \approx n\theta + V_n^*,$$

where

$$V_n^* = e^{-n\theta} W_n.$$

Here V_n^* has a second moment which goes to zero at least as fast as $e^{-2n\theta}$. This suggests that we estimate θ using weighted least squares. The weights should ideally be chosen equal to the variance of $V_n (\approx e^{-2n\theta})$. Since θ is not known, we settle for a conservative (over-) estimate of the variance, $e^{-2n\xi_1}$, and estimate θ by

$$t_{n+1} = \sum_{k=1}^{n} k e^{2k\xi_1} Y_k^* \Big/ \sum_{k=1}^{n} k^2 e^{2k\xi_1}.$$

Here t_{n+1} is related to t_n by the recursion

$$t_{n+1} = t_n + \left(n e^{2n\xi_1} \Big/ \sum_{k=1}^{n} k^2 e^{2k\xi_1} \right) [Y_n^* - n t_n] \qquad (t_1 = 0).$$

If the residuals, W_n, are independent and identically distributed with a density f and have the property

$$\lim_{x \to \pm\infty} \sup |x|^{3+\delta} f(x) < \infty$$

for some positive δ, then it can be rigorously shown that $\mathscr{E}(t_n - \theta)^2 \to 0$ (exponentially fast) as $n \to \infty$.

3. Moment Convergence Rates

We are now going to investigate more closely the large-sample behavior of our estimates $\{t_n\}$ generated by Equation 2.2 when (a) the errors of observation are independent with zero means and (b) some member of the class of gains considered in Theorem 2.3, Condition 3, is used. In fact, for the balance of our treatment of the scalar-parameter case only such gains will be considered, so we repeat the delineation once and for all as follows.

3.1 Restricted Gain Sequence

By a restricted gain sequence we mean a function $a_n(\cdot)$ defined for all points $\mathbf{x} = (x_1, \cdots, x_n)$ in the n-fold product space $J^{(n)}$ of the interval J with itself such that (see Assumption A1)

$$\operatorname{sgn} a_n(\mathbf{x}) = \dot{F}_n$$

for all $\mathbf{x} \in J^{(n)}$, and (see Assumption A3)

$$\inf_{\mathbf{x} \in J^{(n)}} \frac{B_n^2}{b_n} |a_n(\mathbf{x})| \geq a > 0,$$

$$\sup_{\mathbf{x} \in J^{(n)}} \frac{B_n^2}{b_n} |a_n(\mathbf{x})| \leq a' < \infty$$

for some numbers $a \leq a'$ and all sufficiently large n.

As already indicated in Chapter 2, the gains used in practice have this property.

Our first result tells us that the mean-square error tends to zero as $1/B_n^2$ whenever there is such a constant a which exceeds $\frac{1}{2}$, that is, when

$$\liminf_{n} \inf_{\mathbf{x} \in J^{(n)}} \frac{B_n^2 |a_n(\mathbf{x})|}{b_n} > \tfrac{1}{2}.$$

The conditions of Theorem 3.1 are the same as those that ensured strong and mean-square consistency in Theorem 2.3, Condition 3.

3.2 Theorems Concerning Moment Convergence Rates

THEOREM 3.1.

Suppose $\{Y_n : n = 1, 2, \cdots\}$ satisfies Assumptions A1′, A2, A3, A4, and A5″, where $J = (\xi_1, \xi_2)$ is any interval known to contain θ. Furthermore, suppose

$$\sup_n \mathscr{E}[Y_n - F_n(\theta)]^{2q} < \infty$$

for some positive integer q. Let t_1 be arbitrary and

$$t_{n+1} = [t_n + a_n(t_1, \cdots, t_n)[Y_n - F_n(t_n)]]_{\xi_1}^{\xi_2} \qquad (n = 1, 2, \cdots),$$

where $\{a_n\}$ is any restricted gain sequence. Then, if $a > \frac{1}{2}$, it follows that

$$\mathscr{E}(t_n - \theta)^{2q} = O\left(\frac{1}{B_n^{2q}}\right)$$

as $n \to \infty$.

Proof. We let

$$a_n = a_n(t_1, \cdots, t_n), \qquad T_n = t_n + a_n[Y_n - F_n(t_n)]$$

(and note that this meaning of T_n differs from that of Chapter 2). Since the value of t_{n+1} is that of T_n truncated to the interval J, which contains θ, the former must be closer to θ than the latter. Consequently,

$$(t_{n+1} - \theta)^{2p} \leq (T_n - \theta)^{2p} \tag{3.1}$$

for any integer p.

We first derive an upper bound for $\mathscr{E}(T_n - \theta)^{2p}$ which is linear in $\mathscr{E}(t_n - \theta)^{2p}$. By the law of the mean

$$\begin{aligned} T_n - \theta &= t_n - \theta + a_n[F_n(\theta) - F_n(t_n)] + a_n W_n \\ &= [1 - a_n \dot{F}_n(u_n)](t_n - \theta) + a_n W_n, \end{aligned} \tag{3.2}$$

where u_n lies between t_n and θ, and

$$W_n = Y_n - F_n(\theta).$$

Thus,

$$
\begin{aligned}
(T_n - \theta)^{2p} = {} & [1 - a_n\dot{F}_n(u_n)]^{2p}(t_n - \theta)^{2p} \\
& + 2p[1 - a_n\dot{F}_n(u_n)]^{2p-1}(t_n - \theta)^{2p-1}a_nW_n \\
& + \sum_{i=2}^{2p}\binom{2p}{i}[1 - a_n\dot{F}_n(u_n)]^{2p-i}(t_n - \theta)^{2p-i}(a_nW_n)^i.
\end{aligned}
$$

Conditioning by $t_1, t_2, \cdots, t_n$ is tantamount to conditioning by $t_1, W_1, \cdots, W_{n-1}$. Since the zero-mean W's are presumed independent, the second term on the right-hand side has zero conditional expectation, giving

$$
\begin{aligned}
&\mathscr{E}\{(T_n - \theta)^{2p}|t_1, \cdots, t_n\} \\
&\quad = \left[1 - 2pa_n\dot{F}_n(u_n) + \sum_{i=2}^{2p}\binom{2p}{i}[-a_n\dot{F}_n(u_n)]^i\right](t_n - \theta)^{2p} \\
&\quad + \sum_{i=2}^{2p}\binom{2p}{i}[1 - a_n\dot{F}_n(u_n)]^{2p-i}(t_n - \theta)^{2p-i}a_n^i\mathscr{E}W_n^i \\
&\qquad\qquad (p = 1, 2, \cdots, q). \quad (3.3)
\end{aligned}
$$

We now utilize the sure bounds imposed by our assumptions. To save space we set

$$\beta_n = b_n^2/B_n^2, \tag{3.4}$$

an abbreviation which will be used throughout this chapter and the next. Since $a_n\dot{F}_n(u_n) = |a_n|\,|\dot{F}_n(u_n)|$, we then have

$$a\beta_n \le a_n\dot{F}_n(u_n) \le a''\beta_n, \tag{3.5}$$

where

$$a'' = a' \sup d_n/b_n < \infty,$$

the reason being that, because t_n and θ belong to J, u_n must also. The Inequality 3.5 will be valid only for all n exceeding some finite index, which generally depends on a and a'. However, without loss of generality, we may proceed as though the gain restriction is met for $n = 1, 2, \cdots$, and thereby obviate continual rewriting of the qualification. With this understanding, we now majorize the right-hand side of Equation 3.3 by bounding $2pa_n\dot{F}_n(u_n)$ from below and everything else from above with the deterministic quantities in Equation 3.5. Following this, we take expectations and use the sure inequality, Equation 3.1. The result is

$$
\begin{aligned}
\mathscr{E}(t_{n+1} - \theta)^{2p} \le {} & (1 - 2pa\beta_n + K\beta_n^2)\mathscr{E}(t_n - \theta)^{2p} \\
& + K'\sum_{i=2}^{2p}\left(\frac{b_n}{B_n^2}\right)^i\mathscr{E}|t_n - \theta|^{2p-i} \\
& \qquad (p = 1, 2, \cdots, q), \quad (3.6)
\end{aligned}
$$

for all n. Here K and K' are some finite constants depending on p, but not on n, and the latter contains the hypothesized uniform bounds on the observational error moments.

Inequality 3.6 is the starting point in the derivation of moment-convergence rates.

For the presently hypothesized case $a > \frac{1}{2}$, we introduce the working definition

$$X_n' = B_{n-1}(t_n - \theta)$$

and multiply Equation 3.6 through by B_n^{2p}. We get

$$\mathscr{E}X_{n+1}^{2p} \leq \left(\frac{B_n}{B_{n-1}}\right)^{2p} (1 - 2pa\beta_n + K\beta_n^2)\mathscr{E}X_n^{2p} + K' \sum_{i=2}^{2p} \left(\frac{B_n}{B_{n-1}}\right)^{2p-i} \beta_n^{i/2}\, \mathscr{E}|X_n|^{2p-i}. \quad (3.7)$$

Since $\beta_n \to 0$ as $n \to \infty$, we find that

$$\left(\frac{B_n^2}{B_{n-1}^2}\right)^p = \frac{1}{(1 - \beta_n)^p} = 1 + p\beta_n + O(\beta_n^2),$$

where all order relations will be as $n \to \infty$. Thus, for some $c > 0$ and all large enough n, we have

$$\left(\frac{B_n}{B_{n-1}}\right)^{2p} (1 - 2pa\beta_n + K\beta_n^2) = 1 - p(2a - 1)\beta_n + O(\beta_n^2) \leq 1 - c\beta_n, \quad (3.8)$$

because $2a - 1 > 0$. Let N be fixed large enough so that Equation 3.8 holds with $c\beta_n < 1$ for all $n \geq N$. Introduce the inductive hypothesis

$$\mathscr{H}_{p-1}\colon \mathscr{E}X_n^{2(p-1)} = O(1). \quad (3.9)$$

It is *a fortiori* true that the expectations in Equation 3.7 remain bounded as $n \to \infty$ for each index i. Since $\beta_n \to 0$, the summands for $i > 2$ are evidently each of smaller order than the $(i = 2)$-term. Thus, after substituting Equation 3.8 into Equation 3.7, we have

$$\mathscr{E}X_{n+1}^{2p} \leq (1 - c\beta_n)\mathscr{E}X_n^{2p} + K''\beta_n \qquad (\text{all } n \geq N).$$

Iterating back in n to N, we obtain

$$\mathscr{E}X_{n+1}^{2p} \leq \prod_{j=N}^{n} (1 - c\beta_j)\mathscr{E}X_N^{2p} + K'' \sum_{k=N}^{n} \prod_{j=k+1}^{n} (1 - c\beta_j)\beta_k.$$

From the scalar version of the identity which is Lemma 1, it follows that the right-hand side is equal to

$$Q_n\mathscr{E}X_N^{2p} + \frac{K''}{c}(1 - Q_n),$$

where

$$Q_n = \prod_{j=N}^{n} (1 - c\beta_j).$$

Since $\Sigma\beta_n = \infty$ (Equations 3.4 and 2.27 with $r = 0$), Q_n tends to zero as $n \to \infty$. This shows that $\mathscr{H}_{p-1}$ implies $\mathscr{H}_p$. Since $\mathscr{H}_1$ holds trivially and $B_n^2/B_{n-1}^2 \to 1$, the asserted conclusion follows by induction on $p = 1, 2, \cdots, q$. Q.E.D.

For gains with $a \leq \frac{1}{2}$ our technique of estimating convergence rates requires that we strengthen our assumption $\beta_n \to 0$ to $\Sigma\beta_n^2 < \infty$.

THEOREM 3.2

Let the hypotheses of Theorem 3.1 hold with Assumption A5″ strengthened to A5‴. Then

$$\mathscr{E}(t_n - \theta)^{2q} \text{ is at most the order of } \begin{cases} \left(\dfrac{\log B_n^2}{B_n^2}\right)^q & \text{if } a = \frac{1}{2}, \\ \dfrac{1}{B_n^{4qa}} & \text{if } 0 < a < \frac{1}{2}. \end{cases}$$

Proof. We first iterate Inequality 3.6 back to the index N for which

$$z_n = 2pa\beta_n - K\beta_n^2 \in (0, 1) \qquad \text{and} \qquad \log B_n^2 > 1$$

for all $n \geq N$, which can be done since $\beta_n \to 0$ and $B_n^2 \to \infty$. This gives

$$\mathscr{E}(t_{n+1} - \theta)^{2p} \leq \prod_{j=N}^{n} (1 - z_j)\mathscr{E}(t_N - \theta)^{2p} + K' \sum_{k=N}^{n} \prod_{j=k+1}^{n} (1 - z_j) \sum_{i=2}^{2p} \left(\frac{b_k}{B_k^2}\right)^i \mathscr{E}|t_k - \theta|^{2p-i}.$$

We apply Lemma 4 with $z = 2pa$ to get

$$\mathscr{E}(t_{n+1} - \theta)^{2p} \leq \frac{D_{N-1}B_{N-1}^{4pa}\mathscr{E}(t_N - \theta)^{2p}}{B_n^{4pa}} + \frac{K'}{B_n^{4pa}} \sum_{k=N}^{n} D_k \sum_{i=2}^{2p} B_k^{4pa-i}\beta_k^{i/2}\mathscr{E}|t_k - \theta|^{2p-i} \qquad (a > 0), \quad (3.10)$$

where the D_k's are uniformly bounded in accordance with the lemma.

Consider first the case $a = \frac{1}{2}$, and set

$$X_n = \frac{B_{n-1}}{\sqrt{\log B_{n-1}^2}} (t_n - \theta).$$

Multiplication of Equation 3.10 through by $B_n^{2p}/(\log B_n^2)^p$ gives

$$\mathscr{E}X_{n+1}^{2p} \leq \frac{D_{N-1}(\log B_{N-1}^2)^p \mathscr{E}X_N^{2p}}{(\log B_n^2)^p}$$

$$+ \frac{K'}{(\log B_n^2)^p} \sum_{k=N}^{n} D_k \sum_{i=2}^{2p} B_k^{2p-i} \beta_k^{i/2} \mathscr{E}|t_k - \theta|^{2p-i}$$

$$= o(1) + \frac{K'}{(\log B_n^2)^p} \sum_{k=N}^{n} D_k \sum_{i=2}^{2p} \left(\frac{B_k^2 \log B_{k-1}^2}{B_{k-1}^2 \log B_k^2}\right)^{p-i/2}$$

$$\times (\log B_k^2)^{p-i/2} \beta_k^{i/2} \mathscr{E}|X_k|^{2p-i} \quad (3.11)$$

as $n \to \infty$ for all $p \leq q$. We again make the inductive hypothesis, Equation 3.9, but with X_n redefined as in this paragraph. Then the first and fourth factors of the i-summands in Equation 3.11 are bounded uniformly in $k \geq N$. The second factor is never less than unity while the third always is. The $(i = 2)$-term therefore dominates, and

$$\mathscr{E}X_{n+1}^{2p} \leq o(1) + \frac{K''}{(\log B_n^2)^p} \sum_{k=N}^{n} (\log B_k^2)^{p-1} \beta_k$$

$$\leq o(1) + \frac{K''}{\log B_n^2} \sum_{k=N}^{n} \beta_k,$$

because $\log B_k^2$ increases with k. But the last written sum is $O(1)$ as $n \to \infty$ by Equation L4.3 of the Appendix. In fact, we have (Knopp, 1947, p. 292),

$$\sum_{k=1}^{n} \frac{b_k^2}{B_k^2} \cong \log B_n^2,$$

which, incidentally, makes explicit the Abel–Dini Theorem (2.27) for $r = 0$. (The symbol $\cong$ will always mean the ratio of the two sides tends to unity.) Therefore, $\mathscr{H}_{p-1}$ implies $\mathscr{H}_p$, and the proof for $a = \frac{1}{2}$ is completed by induction on $p = 1, 2, \cdots, q$.

For the remaining case, we redefine the working variate as

$$X_n = B_{n-1}^{2a}(t_n - \theta) \qquad (0 < a < \tfrac{1}{2})$$

and multiply Equation 3.10 through by B_n^{4pa}. In place of the final bound in Equation 3.11, we find

$$\mathscr{E}X_{n+1}^{2p} \leq D_{N-1}\mathscr{E}X_N^{2p} + K' \sum_{k=N}^{n} D_k$$

$$\times \sum_{i=2}^{2p} \left(\frac{B_k^2}{B_{k-1}^2}\right)^{a(2p-i)} \left(\frac{\beta_k^2}{B_k^{2-4a}}\right)^{i/2} \mathscr{E}|X_k|^{2p-i}.$$

As before, the sum on i is seen to be bounded by a constant independent of k times the ($i = 2$)-term, so that under the inductive hypothesis $\mathscr{H}_{p-1}$ we have

$$\mathscr{E}X_{n+1}^{2p} \le O(1) + K'' \sum_{k=N}^{n} \frac{\beta_k}{B_k^{2-4a}}.$$

The last written sum is the same thing as

$$\sum_{k=N}^{n} \frac{b_k^2}{B_k^{2+2(1-2a)}},$$

which, according to Equation 2.27, tends to a finite limit as $n \to \infty$. Therefore, $\mathscr{H}_p$ holds, and the usual induction finishes the argument. Q.E.D.

The conclusions of Theorems 3.1 and 3.2 depend on the value of a that results from our selection of a gain sequence. Letting

$$c_n = \inf_{\mathbf{x} \in J^{(n)}} \frac{B_n^2|a_n(\mathbf{x})|}{b_n},$$

we see that there always exists an $a > \frac{1}{2}$ when $L = \lim_n \inf c_n > \frac{1}{2}$, and an $a < \frac{1}{2}$ when $L < \frac{1}{2}$. Generally speaking, the case $a = \frac{1}{2}$ occurs only when $L = \frac{1}{2}$ and $c_n < \frac{1}{2}$ for only finitely many values of n.

It is important to note that *the assumption* $a > \frac{1}{2}$ *is a necessary one for the conclusion of Theorem 3.1*, at least when Assumption A5‴ also holds. To demonstrate this, assume the gain restriction is satisfied by some largest (smallest) number a (a'), and that $\text{Var } Y_n \ge \sigma_0^2 > 0$ for all n. Furthermore, let us take $J = (-\infty, \infty)$, which forces equality in Equation 3.1. Then, if we use Equation 3.5 in Equation 3.3 with $p = 1$, we get

$$\mathscr{E}\{(t_{n+1} - \theta)^2 | t_1, \cdots, t_n\} \ge (1 - 2a''\beta_n)(t_n - \theta)^2 + a^2\sigma_0^2 \frac{b_n^2}{B_n^4}$$

after dropping the positive term $a^2\beta_n^2(t_n - \theta)^2$. Taking expectations and iterating back to an appropriate finite index N, we obtain, with e_n^2 again abbreviating $\mathscr{E}(t_n - \theta)^2$,

$$e_{n+1}^2 \ge \prod_{j=N}^{n} (1 - 2a''\beta_j)e_N^2 + a^2\sigma_0^2 \sum_{k=N}^{n} \prod_{j=k+1}^{n} (1 - 2a''\beta_j) \frac{b_k^2}{B_k^4}.$$

If we assume $\Sigma\beta_n^2 < \infty$, we can apply Lemma 4 with $z = 2a''$ and $K = 0$. Thus, after further weakening our lower bound by dropping the positive term involving e_N^2, we have

$$e_{n+1}^2 \ge \frac{a^2\sigma_0^2}{B_n^{4a''}} \sum_{k=N}^{n} C_k \frac{b_k^2}{B_k^{4-4a''}}$$

$$\ge \frac{\text{const}}{B_n^{4a''}} \sum_{k=N}^{n} \frac{b_k^2}{B_k^{2+2(1-2a'')}}.$$

The strictly positive "const" involves a uniform lower bound on the C's, which exists according to Lemma 4. Using Equation 2.27 once again, we see that

$$\liminf_n B_n^{4a''} e_n^2 = \begin{cases} +\infty & \text{if } a'' \geq \frac{1}{2}, \\ A > 0 & \text{if } a'' < \frac{1}{2}. \end{cases}$$

Thus, if the assumption $a > \frac{1}{2}$ of Theorem 3.1 fails, the mean-square error cannot generally be $O(1/B_n^2)$. Indeed,

$$B_n^2 e_n^2 = B_n^{2(1-2a'')} \cdot B_n^{4a''} e_n^2 \to \infty$$

for all $a'' < \frac{1}{2}$, that is, all cases in which

$$1 \leq \sup_n \frac{d_n}{b_n} < \frac{1}{2a}.$$

3.3 Power–Law Derivatives

We have shown in Theorem 2.4 that the conditions of Theorems 3.1 and 3.2 hold for polynomial regression when the function multiplying the largest power of n is bounded, differentiable, and 1-1 on J. In any such problem the sequence of Equation 3.4 goes to 0 as $1/n$. We point out here that this is true for the larger class of squared infimums of the form

$$b_n^2 = l_n n^\beta \qquad (-1 < \beta < \infty), \tag{3.12}$$

where $l_n > 0$ is any "slowly" increasing or decreasing sequence, i.e., one for which

$$\frac{l_{n+1}}{l_n} = 1 + o\left(\frac{1}{n}\right).$$

In fact, for any such sequence (see, for example, Lemma 4 in Sacks, 1958), we have

$$\beta_n = \frac{b_n^2}{B_n^2} \cong \frac{\beta + 1}{n}.$$

We should not infer from this that $nb_n^2 \to \infty$ is necessary to meet our conditions $B_n^2 \to \infty$ and $\Sigma \beta_n^2 < \infty$. Indeed, if

$$b_n^2 = \frac{1}{n \log n}, \tag{3.13}$$

it is true that $B_n^2 \cong \log \log n$, and hence $\beta_n = o(1/n)$. We retain this β_n behavior, and make B_n^2 increase even more slowly, when we replace

$\log n$ in Equation 3.13 by the product of iterated logarithms (see Knopp, 1947, p. 293).

At the other end of the spectrum, we cannot (as already noted in Chapter 2) handle derivatives which increase faster than some finite power of n, that is, exponentially fast. In such cases, two assumptions are violated because $d_n/b_n \to \infty$ and $b_n^2/B_n^2 \not\to 0$. Although the latter can be compensated for by dividing the gains by $1/n$, the former cannot. An ad hoc treatment like the one used in Section 2.9 is required.

3.4 Relevance to Stochastic Approximation

At the end of Chapter 1, we rewrote our iterative procedure as a Robbins–Monro-type process. Here we pursue this point a bit further and relate the two preceding theorems to some known results in the theory of stochastic approximation. For this purpose we will take J to be the entire real axis.

Consider the following situation. For each real number x and integer n, let $Z_n(x)$ be an observable random variable whose mean value can be written in the form

$$\mathscr{E}Z_n(x) = G_n(x) = K_n(x)(x - \theta) \tag{3.14}$$

for some θ. Furthermore, suppose $\sup_{x,n} \operatorname{Var} Z_n(x) < \infty$. The function $K_n(x)$, which may depend on θ, is assumed one-signed, say $\inf_x K_n(x) > 0$ for every fixed n. Corresponding to a choice of weights $\alpha_n > 0$, we then estimate θ by

$$t_{n+1} = t_n - \alpha_n z_n \qquad (n = 1, 2, \cdots; t_1 \text{ arbitrary}), \tag{3.15}$$

where z_n denotes a random variable whose conditional distribution, given $t_1 = \xi_1, t_2 = \xi_2, \cdots, t_n = \xi_n$, is the distribution of $Z_n(\xi_n)$ (or, equivalently, the conditioning can be on the values of $t_1, z_1, \cdots, z_{n-1}$).

This is Burkholder's (1956) type-A_0 process specialized to the case where the regression functions all have the same zero. The significance of our results lies in their validity for a much larger class of $K_n(x)$'s than heretofore considered.

To apply Theorems 3.1 and 3.2, it clearly suffices, in accordance with Equations 1.8 and 1.9, to make the following symbolic identifications:

$$\begin{aligned} K_n(x) &= |\dot{F}_n(u_n)|, & u_n &= u_n(x, \theta); \\ Z_n(x) &= \operatorname{sgn} \dot{F}_n[F_n(x) - Y_n], & z_n &= Z_n(t_n); \\ \alpha_n &= |a_n|. \end{aligned} \tag{3.16}$$

Independence of the Y_n's is essentially the stated property of the z_n's

Our assumptions A2, A3, A4, and A5‴ place restrictions on

$$b_n = \inf_x K_n(x), \qquad d_n = \sup_x K_n(x). \tag{3.17}$$

Thus, if a_n is chosen as any restricted gain sequence, we have for the mean-square estimation error of the successive approximations (Equation 3.15)

$$\mathscr{E}(t_n - \theta)^2 \text{ is at most the order of} \begin{cases} 1/B_n^2 & \text{for } \frac{1}{2} < a < \infty, \\ \log B_n^2/B_n^2 & \text{for } a = \frac{1}{2}, \\ 1/B_n^{4a} & \text{for } 0 < a < \frac{1}{2}, \end{cases} \tag{3.18}$$

as n tends to infinity. There is, of course, a concealed limitation on applicability: we need to know the n-dependence of the infimums b_n in order to select the proper gain sequence.

In the degenerate case $b_n = b_0 > 0$ and $d_n = d_0 < \infty$ for all n, we have essentially the model under which the original Robbins–Monro process was studied, namely, that an unknown regression function falls between some pair of fixed straight lines passing through the point to be estimated (and hence G_n might as well be viewed as independent of n). Since $B_n^2/b_n = b_0 n$, Assumptions A2, A3, A4, and A5‴ are obviously satisfied. Furthermore, $\alpha_n = \alpha/n$ is a restricted gain sequence, no matter what positive value we fix for α. The resulting Robbins–Monro process is mean-square convergent for all positive values of

$$a = b_0\alpha = \text{minimum slope} \times \text{gain const.}$$

The way in which the rate of convergence is governed by this product is given by Equation 3.18 with B_n^2 proportional to n. This special case has been derived by Hodges and Lehmann (1956).

If we assume that

$$\left.\frac{dG_n(x)}{dx}\right|_{x=\theta} = K \qquad \text{and} \qquad \left.\operatorname{Var} Z_n(x)\right|_{x=\theta} = \sigma^2,$$

for all n, the deviations $\sqrt{n}\,(t_n - \theta)$ of the Robbins–Monro process tend to be normally distributed about the origin in large samples. The variance (Sacks, 1958, Theorem 1) is

$$\frac{\alpha^2\sigma^2}{2K\alpha - 1} \leq \frac{\sigma^2}{b_0^2} V(a), \qquad \text{where} \quad V(a) = \frac{a^2}{2a - 1},$$

provided that α is chosen large enough to make $a > \frac{1}{2}$. Here V is minimized by $a = 1$, that is, by $\alpha = 1/b_0$ (cf. Newton's method with a constant correction weight). As we will see in the next chapter, the function V appears in the limiting distribution of our estimates generated by various choices of the restricted gain sequence.

3.5 Generalization

As emphasized in Chapter 1, we are interested in deriving general results in the theory of stochastic approximation only insofar as they pertain to the analysis of the recursive estimation scheme for nonlinear regression problems. However, it seems appropriate to note here that we could have written Theorems 3.1 and 3.2 not only in the wider formulation of Equations 3.14 through 3.17 but, moreover, with the first of these replaced by

$$G_n(x) = K_n(x)(x - \theta_n).$$

It is not difficult to show that the conclusions hold as written, provided that the roots of the regression functions are such that

$$\theta_n - \theta = o\left(\frac{1}{B_n}\right)$$

as $n \to \infty$.

Theorem 3.1 so generalized is Burkholder's (1956) Theorem 2 (after we ignore the continuous convergence portion of his conditions which are imposed to show that $B_n^{2q}\mathscr{E}(t_n - \theta)^{2q}$ is not only $O(1)$ but tends to the $2q$th moment of a certain normal variate). However, Assumptions A3 and A5″ permit a much larger class of $K_n(x)$'s than does his corresponding assumption that b_n^2 is of the form of Equation 3.12 without the l_n's and the exponent restricted to $-1 < \beta \leq 0$.

4. Asymptotic Distribution Theory

None of the results in Chapter 2 or 3 depended on the nature of the "iteration space" J other than that it should contain the true parameter point. However, when we turn to the question of a limiting distribution for the successive iterates, we will find we need stronger conditions if one of the end points is finite.

Theorem 4.1 assumes that J is the entire real axis. As already pointed out at the beginning of Chapter 2, this still covers cases where θ is known to belong to a finite or semifinite interval, say J_0, and an untruncated procedure is used by linearly extending the regression functions to $(-\infty, \infty)$. On the other hand, when the iterates are confined to such an interval J_0, the hypotheses of Theorem 4.1′ require the existence of higher-order moments. The number of these, over and above the fourth, depends on how fast the "signal" is becoming "buried in the noise," that is, how fast the regression functions are flattening out (if in fact they do).

Theorem 4.1 is not immediately applicable (although it does have theoretical interest in its own right). In fact, it might better be regarded as a lemma for Theorem 4.2, where we show that its subsidiary hypotheses are indeed satisfied by some particular gains. The new assumption concerning the functions $|\dot{F}_n|/b_n$ can be replaced by a different one, which is discussed after the proofs. First, however, we will need the following.

4.1 Notation for and Relations Between Modes of Convergence

The following are (standard) symbolic descriptors for the asymptotic behavior of sequences of random variables. We list them here and use them in this chapter and the next without referring to them again.

1. $X_n = O_p(1)$ means X_n remains bounded in probability as $n \to \infty$.
2. $X_n \xrightarrow{r} X$ means X_n converges to X in the rth mean ($r > 0$).
3. $X_n \overset{\text{a.s.}}{\to} X$ means X_n converges to X with probability one.
4. $X_n \xrightarrow{P} X$ or $X_n = X + o_p(1)$ means X_n converges to X in probability.
5. $X_n \sim Y_n$ means X_n and Y_n have the same limiting distribution. In particular, if $Y_n \sim Y$ and Y is normal with mean 0 and variance ψ^2, we write $X_n \sim N(0, \psi^2)$.

Certain mutual implications will frequently be used: Mode 4 is a consequence of either 2 or 3; Mode 4 and a.s. $\sup_n |X_n| < \infty$ imply Mode 2; and 5 implies 1 (see Loève, 1960, Chapter 3). Furthermore, if $X_n \sim X$ and $Y_n = o_p(1)$, then $X_n + Y_n \sim X$, and Mode 4 is preserved through continuous transformations. There is a calculus of o_p and O_p, analogous to o and O; for example, $o_p(1)O_p(1) = o_p(1)$ (see Chernoff, 1956).

4.2 Theorems Concerning Asymptotic Normality for General Gains

THEOREM 4.1

Suppose that $\{Y_n : n = 1, 2, \cdots\}$ satisfies Assumptions A1′, A2, A3, A4, and A5″ with $J = (-\infty, \infty)$, and set

$$g_n(x) = \frac{|\dot{F}_n(x)|}{b_n}.$$

Suppose the functions $g_1, g_2, \cdots$ are continuously convergent at the point θ; that is to say, for every sequence $\{x_n\}$ tending to θ as $n \to \infty$, $\{g_n(x_n)\}$ has a limit. Furthermore, suppose that

$$\operatorname{Var} Y_n = \sigma^2 \qquad \text{and} \qquad \sup_n \mathscr{E}|Y_n - F_n(\theta)|^{2+\delta} < \infty,$$

for some $\delta > 0$. Let t_1 be the arbitrary and

$$t_{n+1} = t_n + a_n(t_1, \cdots, t_n)[Y_n - F_n(t_n)] \qquad (n = 1, 2, \cdots),$$

where $\{a_n\}$ is a restricted gain sequence such that

$$L = \liminf_n \inf_{\mathbf{x} \in J^{(n)}} \frac{B_n^2}{b_n} |a_n(\mathbf{x})| > \tfrac{1}{2}.$$

Then

$$B_n(t_n - \theta) \sim N\left(0, \frac{\sigma^2\mu}{2\lambda - 1}\right)$$

as $n \to \infty$, provided that the function sequences $\{g_n\}$ and $\{a_n\}$ satisfy the following further conditions:

1. $\lim_n \mathscr{E}\left[\frac{B_n^2}{b_n}|a_n(t_1, \cdots, t_n)|\, g_n(t_n) - \lambda\right]^2 = 0,$

2. $\lim_n \mathscr{E}\left[\frac{B_n^4}{b_n^2} a_n^2(t_1, \cdots, t_n) - \mu\right]^2 = 0,$

for some $\lambda > \frac{1}{2}$ and $\mu > 0$.

Proof. We introduce the abbreviations

$$y_n = \frac{B_n^2}{b_n}|a_n(t_1, \cdots, t_n)|\, g_n(t_n),$$

$$y_n' = \frac{B_n^2}{b_n}|a_n(t_1, \cdots, t_n)|\, g_n(u_n) \qquad (|u_n - \theta| < |t_n - \theta|), \tag{4.1}$$

where u_n is the point, with the indicated property, which arose in Equation 3.2 from the law of the mean. Assumption concerning the (bounded) random variable y_n is made under Condition 1 in the statement of the theorem to be proved. Letting λ be as it is postulated there, we rewrite our untruncated difference equation

$$t_{n+1} - \theta = (1 - |a_n\dot{F}_n(u_n)|)(t_n - \theta) + a_n W_n$$

as

$$\begin{aligned} t_{n+1} - \theta &= (1 - y_n'\beta_n)(t_n - \theta) + a_n W_n \\ &= (1 - \lambda\beta_n)(t_n - \theta) - \beta_n[(y_n' - y_n) + (y_n - \lambda)] \\ &\qquad \times (t_n - \theta) + a_n W_n, \end{aligned} \tag{4.2}$$

where a_n and W_n are the same as in Equation 3.2 and β_n is still given by Equation 3.4. After iterating this back to an arbitrary fixed integer N, we obtain

$$\begin{aligned} t_{n+1} - \theta &= \prod_{j=N}^{n} (1 - \lambda\beta_j)(t_N - \theta) \\ &\quad - \sum_{k=N}^{n} \prod_{j=k+1}^{n} (1 - \lambda\beta_j)[\beta_k(y_k' - y_k)(t_k - \theta) \\ &\quad + \beta_k(y_k - \lambda)(t_k - \theta) - a_k W_k] \\ &= \mathrm{I} + \mathrm{II} + \mathrm{III} + \mathrm{IV}. \end{aligned} \tag{4.3}$$

We are going to show, under the conditions of the theorem, that I, II, and III go to zero in first mean faster than $1/B_n$, while B_n times IV has the asserted limiting normal distribution. We fix N, as usual, sufficiently large so that $\lambda\beta_j < 1$ for all $j \geq N$.

With regard to I: From Lemma 4, we have

$$\prod_{j=k+1}^{n} (1 - \lambda\beta_j) \leq \frac{D_k B_k^{2\lambda}}{B_n^{2\lambda}} = o\left(\frac{1}{B_n}\right), \tag{4.4}$$

for any fixed k (in particular, $k = N - 1$), because 2λ exceeds one by assumption.

With regard to II: Using Equation 4.4 and the Schwarz Inequality, we obtain

$$B_n\mathscr{E}|\text{II}| \leq \sum_{k=N}^{n} D_k \frac{B_k^{2\lambda-1}\beta_k}{B_n^{2\lambda-1}} \cdot \mathscr{E}^{1/2}(y_k' - y_k)^2 B_k e_k. \tag{4.5}$$

According to the hypothesis of the theorem, we have $L \geq \frac{1}{2} + 2\varepsilon$ for some $\varepsilon > 0$. By definition of the limit inferior there corresponds, to any such ε, a finite index n_ε such that, for all $n > n_\varepsilon$,

$$\inf_{\mathbf{x} \in J(n)} \frac{B_n^2}{b_n} |a_n(\mathbf{x})| \geq L - \varepsilon \geq \frac{1}{2} + \varepsilon.$$

The gain restriction is therefore satisfied by some number $a = \frac{1}{2} + \varepsilon > \frac{1}{2}$, so that by Theorem 3.1

$$e_k = \mathscr{E}^{1/2}(t_k - \theta)^2 = O\left(\frac{1}{B_k}\right)$$

as $k \to \infty$, independent of the value of a. Next, let γ denote the hypothesized limiting value of $g_k(x_k)$ when x_k tends to θ. Then from Equation 4.1 and the gain restriction,

$$|y_k' - y_k| \leq a'(|g_k(u_k) - \gamma| + |g_k(t_k) - \gamma|) = v_k \quad \text{(say)}. \tag{4.6}$$

Since t_k, and hence u_k, converges in probability to θ, it follows that $v_k = o_p(1)$ as $k \to \infty$. But for all n and x, we have

$$1 \leq g_n(x) \leq c_0 \qquad \left(c_0 = \sup_n \frac{d_n}{b_n} < \infty\right), \tag{4.7}$$

and therefore the v_k's are bounded random variables and $\mathscr{E}v_k^2 \to 0$. The sequence following the center dot in Equation 4.5 is thus $o(1)O(1) = o(1)$ as $k \to \infty$. By Lemma 4, $\sup_{k \geq N} D_k < \infty$. Therefore, since $2\lambda - 1 > 0$, the bound in Equation 4.5 must go to zero as $n \to \infty$ by Lemma 5.

With regard to III: In the same way, we find that

$$B_n\mathscr{E}|\text{III}| \leq \text{const} \sum_{k=N}^{n} \frac{B_k^{2\lambda-1}\beta_k}{B_n^{2\lambda-1}} \cdot \mathscr{E}^{1/2}(y_k - \lambda)^2 B_k e_k.$$

By Condition 1, Theorem 3.1, and Lemma 5, this bound also goes to zero as $n \to \infty$.

With regard to IV: The preceding and Equation 4.3 combine to give

$$B_n(t_n - \theta) \sim \sum_{k=N}^{n} B_n \prod_{j=k+1}^{n} (1 - \lambda\beta_j)a_k W_k \equiv X_n$$

as $n \to \infty$. To show that X_n has the asserted large-sample normal distribution, we express this sum in the formulation of Lindeberg's Central Limit Theorem (Loève, 1960, p. 377):

$$X_n = \sum_{k=1}^{n} X_{nk}, \qquad X_{nk} = a_{nk}W_k,$$

$$a_{nk} = \begin{cases} 0 & \text{for } k = 1, 2, \cdots, N-1, \\ B_n \prod_{j=k+1}^{n} (1 - \lambda\beta_j)a_k & \text{for } k = N, N+1, \cdots, n. \end{cases} \tag{4.8}$$

The multipliers a_{nk} are random variables via

$$a_k = a_k(t_1, \cdots, t_k).$$

From the form of the iteration, it is clear that

$$\left.\begin{array}{l} t_1, t_2, \cdots, t_k \\ t_1, W_1, \cdots, W_{k-1} \\ t_1, X_{n1}, \cdots, X_{n,k-1} \end{array}\right\} \text{are equivalent conditioning sets.} \tag{4.9}$$

Thus,

$$\mathscr{E}X_{nk} = \mathscr{E}\mathscr{E}\{X_{nk}|t_1, t_2, \cdots, t_k\} = \mathscr{E}a_{nk}\mathscr{E}\{W_k|t_1, W_1, \cdots, W_{k-1}\} = 0$$

by the assumed independence of the errors $W_k = Y_k - F_k(\theta)$. The summands are therefore centered.

Next, we set

$$\chi_{nk}(\varepsilon) = \begin{cases} 1 & \text{if } |X_{nk}| > \varepsilon \\ 0 & \text{otherwise,} \end{cases}$$

where $\varepsilon > 0$ is arbitrary, and

$$\hat{\sigma}_{nk}^2 = \mathscr{E}\{X_{nk}^2|t_1, X_{n1}, \cdots, X_{n,k-1}\}$$

$$\sigma_{nk}^2 = \mathscr{E}X_{nk}^2.$$

A special case of Lindeberg's theorem tells us that

$$X_n \sim N(0, \psi^2)$$

if the following conditions hold:

$$\text{a. } \lim_n \sum_{k=1}^n \mathscr{E}\chi_{nk}(\varepsilon)X_{nk}^2 = 0,$$

$$\text{b. } \lim_n \sum_{k=1}^n \mathscr{E}|\hat{\sigma}_{nk}^2 - \sigma_{nk}^2| = 0, \tag{4.10}$$

$$\text{c. } \lim_n \sum_{k=1}^n \sigma_{nk}^2 = \psi^2 < \infty.$$

It remains to prove that Equation 4.10 is a consequence of our assumptions, with the asserted formula for ψ^2.

With regard to a: By Hölder's Inequality, we have

$$\mathscr{E}\chi_{nk}(\varepsilon)X_{nk}^2 \leq P^{\delta/(1+\delta)}\{|a_{nk}W_k| > \varepsilon\}\mathscr{E}^{1/(1+\delta)}a_{nk}^{2(1+\delta)}W_k^{2(1+\delta)},$$

where 2δ is the δ of the theorem's hypothesis. From Equations 4.4, 4.8, and the gain restriction, we obtain a sure bound

$$|a_{nk}| \leq \text{const}\,\frac{B_k^{2\lambda-1}}{B_n^{2\lambda-1}}\frac{b_k}{B_k} = \alpha_{nk} \qquad \text{(say)}.$$

By Markov's Inequality, we have

$$P\{|a_{nk}W_k| > \varepsilon\} \leq \frac{\alpha_{nk}^{2(1+\delta)}\mathscr{E}W_k^{2(1+\delta)}}{\varepsilon^{2(1+\delta)}}.$$

Thus, we obtain

$$\mathscr{E}\chi_{nk}(\varepsilon)X_{nk}^2 \leq \frac{\alpha_{nk}^{2(1+\delta)}\mathscr{E}W_k^{2(1+\delta)}}{\varepsilon^{2\delta}} \leq \frac{\text{const}}{\varepsilon^{2\delta}}\frac{B_k^{2(2\lambda-1)(1+\delta)}\beta_k}{B_n^{2(2\lambda-1)(1+\delta)}}\cdot\beta_k^{\delta}.$$

Condition a follows from Lemma 5 because $2\lambda - 1 > 0$ and $\lim_{k\to\infty}\beta_k^{\delta} = 0$.

With regard to b: In addition to Equation 4.1, we need one more abbreviation; namely,

$$z_n = \frac{B_n^4}{b_n^2}a_n^2(t_1, \cdots, t_n) = \frac{B_n^2}{\beta_n}a_n^2(t_1, \cdots, t_n). \tag{4.11}$$

According to Equation 4.9,

$$\hat{\sigma}_{nk}^2 = \begin{cases} 0, & k = 1, 2, \cdots, N-1, \\ \sigma^2 B_n^2 \prod_{j=k+1}^n (1-\lambda\beta_j)^2 \dfrac{\beta_k}{B_k^2} z_k, & k = N, N+1, \cdots, n. \end{cases} \tag{4.12}$$

Thus, from Equation 4.4, we obtain

$$\mathscr{E}|\hat{\sigma}_{nk}^2 - \sigma_{nk}^2| \leq D_k^2 \frac{B_k^{2(2\lambda-1)}\beta_k}{B_n^{2(2\lambda-1)}} \cdot \mathscr{E}|z_k - \mathscr{E}z_k|, \tag{4.13}$$

because $\sigma_{nk}^2 = \mathscr{E}\hat{\sigma}_{nk}^2$ (the left-hand side, of course, being zero for $k < N$). But

$$\mathscr{E}|z_k - \mathscr{E}z_k| \leq \mathscr{E}|z_k - \mu| + |\mathscr{E}(z_k - \mu)| \leq 2\mathscr{E}^{1/2}(z_k - \mu)^2. \tag{4.14}$$

After substituting Equation 4.14 into Equation 4.13 and summing over k from 1 to N, Condition b follows from Condition 2 and Lemma 5.

With regard to c: We have from Equation 4.12, in the notation of Lemma 6,

$$\begin{aligned}\sum_{k=1}^{n} \sigma_{nk}^2 &= \sigma^2 B_n^2 \sum_{k=N}^{n} \prod_{j=k+1}^{n} (1 - \lambda\beta_j)^2 \frac{\beta_k}{B_k^2} [\mu + \mathscr{E}(z_k - \mu)] \\ &= \sigma^2 \mu \Psi_n^2(\lambda) + o(1),\end{aligned}$$

where the order term exists for reasons already given in the previous paragraph. We immediately obtain

$$\psi^2 = \frac{\sigma^2\mu}{2\lambda - 1}$$

by the conclusion of Lemma 6. Q.E.D.

Remark. The restriction to gains for which there exists an $a > \frac{1}{2}$ guarantees that $e_n = O(1/B_n)$ and, consequently, under the present assumptions, that II and III in Equation 4.3 are both $o_p(1/B_n)$. As pointed out following the proof of Theorem 3.2, it is necessary, when Assumption A5‴ holds, to have $a > \frac{1}{2}$ to ensure this rate of convergence. Assumption A5‴, in turn, was needed to apply Lemma 6 and get a definite limit ψ^2 in Equation 4.10.

THEOREM 4.1′

Let the hypotheses of Theorem 4.1 hold over $J = [\xi_1, \xi_2]$, with at least one of the end points finite, and suppose we choose the interval so that θ is an interior point. In addition, assume there exists an integer p, $2 \leq p < \infty$, for which

$$\lim_n B_n^{p-2} b_n = \infty$$

and, corresponding to the smallest such integer,

$$\sup_n \mathscr{E}[Y_n - F_n(\theta)]^{2p} < \infty.$$

Let t_1 be arbitrary, and

$$t_{n+1} = [t_n + a_n(t_1, \cdots, t_n)[Y_n - F_n(t_n)]]_{\xi_1}^{\xi_2} \qquad (n = 1, 2, \cdots),$$

where $\{a_n\}$ is any restricted gain sequence having $L > \frac{1}{2}$. Then the conclusion of Theorem 4.1 holds under Conditions 1 and 2.

Proof. We represent the effect of truncation as an additional term on the right-hand side of the fundamental formula, Equation 4.3, in the following way. Let T_n again abbreviate the function $t_n + a_n[Y_n - F_n(t_n)]$. Define the indicators

$$\chi_{n1} = \begin{cases} 1 & \text{if } T_n \le \xi_1, \\ 0 & \text{otherwise,} \end{cases} \qquad \chi_{n2} = \begin{cases} 1 & \text{if } T_n \ge \xi_2, \\ 0 & \text{otherwise,} \end{cases}$$

and the random variable

$$U_n = \chi_{n1}\xi_1 + \chi_{n2}\xi_2 - (\chi_{n1} + \chi_{n2})T_n. \tag{4.15}$$

In what follows we proceed as though both end points are finite. If one is not, the appropriate term in U_n is to be deleted and the ensuing arguments accordingly modified.

In this notation the truncated recursion is

$$t_{n+1} = [T_n]_{\xi_1}^{\xi_2} = T_n + U_n,$$

and $T_n - \theta$ is given by the right-hand side of Equation 4.2:

$$T_n - \theta = (1 - y_n'\beta_n)(t_n - \theta) + a_n W_n. \tag{4.16}$$

The meaning of all symbols is the same as before, the only difference being that $t_1, \cdots, t_n, u_n$ and θ now belong to a finite interval. We thus have

$$t_{n+1} - \theta = (\text{right-hand side of Equation 4.3}) + \bar{U}_n,$$

$$\bar{U}_n = \sum_{k=N}^{n} \prod_{j=k+1}^{n} (1 - \lambda\beta_j)U_k.$$

The hypotheses of the present theorem include those of Theorem 4.1, after the latter is rewritten for a finite interval. The conclusion will thus be at hand once we show that $B_n\bar{U}_n = o_p(1)$.

From Equation 4.4, since $\sup_{k \ge N} D_k < \infty$, we have

$$B_n|\bar{U}_n| \le \text{const} \sum_{k=N}^{n} \frac{B_k^{2\lambda-1}\beta_k}{B_n^{2\lambda-1}} \cdot \frac{B_k}{\beta_k}|U_k|.$$

After taking expectations and using Lemma 5, we see that it suffices to prove the stronger statement

$$\mathscr{E}|U_n| = o\left(\frac{\beta_n}{B_n}\right) \tag{4.17}$$

as $n \to \infty$, and this is what we now proceed to do.

If we set

$$\xi = \max(|\xi_1|, |\xi_2|) < \infty \qquad \text{and} \qquad \chi_n = \chi_{n1} + \chi_{n2},$$

then, from Equation 4.15, we have

$$|U_n| \le \xi\chi_n + |T_n|\chi_n.$$

All quantities on the right-hand side of Equation 4.16 are surely bounded, with the possible exception of $W_n = Y_n - F_n(\theta)$. Therefore, $|T_n|$ has as many moments as $|W_n|$, which by hypothesis is $2p$. From the Hölder Inequality and the fact that $\mathscr{E}^{1/r}|X|^r$ is a nondecreasing function of real numbers r, it follows that

$$\mathscr{E}|U_n| \le \xi\mathscr{E}\chi_n + \mathscr{E}^{1/2p}|T_n|^{2p} \cdot \mathscr{E}^{(2p-1)/2p}\chi_n^{2p/(2p-1)}$$
$$\le \text{const}\, \mathscr{E}^{(2p-1)/2p}\chi_n. \tag{4.18}$$

We seek, therefore, the n-dependence of $\mathscr{E}\chi_n$.

The random variable y_n' in Equation 4.16 belongs to a finite interval of positive numbers (see Equations 4.1 and 4.7). Hence, for all large enough n, Equation 4.16 gives

$$|T_n - \theta| \le |t_n - \theta| + a'\frac{b_n}{B_n^2}|W_n|,$$

because β_n tends to zero. Since we are assuming, without loss of generality, that $\xi_1 < \theta < \xi_2$, we can write

$$\theta - \xi_1 \ge 2\varepsilon > 0, \qquad \xi_2 - \theta \ge 2\varepsilon > 0,$$

for some such ε. For the right-hand end point, we therefore have

$$\mathscr{E}\chi_{n2} = P\{T_n - \theta \ge \xi_2 - \theta\} \le P\{|T_n - \theta| \ge 2\varepsilon\}$$
$$\le P\{|t_n - \theta| \ge \varepsilon\} + P\left\{|W_n| \ge \frac{\varepsilon B_n^2}{a'b_n}\right\}$$
$$\le \frac{\mathscr{E}(t_n - \theta)^{2p}}{\varepsilon^{2p}} + \left(\frac{a'b_n}{\varepsilon B_n^2}\right)^{2p}\mathscr{E}|W_n|^{2p}$$

after using Markov's Inequality. The second term, in the notation of Equation 3.4, is $O(\beta_n{}^p/B_n{}^{2p}) = o(1/B_n{}^{2p})$ and hence, according to Theorem 3.1, of smaller order than the first. It is clear that $\mathscr{E}\chi_{n1}$ can be bounded by a sequence with the same n-dependence; hence

$$\mathscr{E}\chi_n = O\left(\frac{1}{B_n{}^{2p}}\right).$$

Consequently, from Equation 4.18,

$$\mathscr{E}|U_n| = O\left(\frac{1}{B_n{}^{2p-1}}\right).$$

But

$$\frac{1}{B_n{}^{2p-1}} = \frac{1}{(B_n{}^{p-2}b_n)^2} \cdot \frac{\beta_n}{B_n} = o\left(\frac{\beta_n}{B_n}\right)$$

by hypothesis. This establishes Equation 4.17 which, as already argued, is sufficient. Q.E.D.

Remark. The assumption that $B_n{}^{p-2}b_n \to \infty$ for some finite p is directed at those situations in which the sequence of derivative functions tends to zero as $n \to \infty$, and it places a limitation on the way in which we allow this to happen. It excludes Equation 3.13, and also infimums like $b_n{}^2 = \log n/n$, since then we would have $B_n{}^2 \cong \log^2 n$. However, the assumption is satisfied by Equation 3.12, with $p \geq 2$ the first integer exceeding $2 - \beta/(\beta + 1)$. This makes quantitative the required relationship between the rate at which the derivatives are approaching zero and the number of existing noise moments.

4.3 Alternative to the Continuous Convergence Assumption

Referring to the proof of Theorem 4.1, we see that the assumption of continuous convergence of $\{g_n(\cdot)\}$ at the point θ was explicitly used to show that the expectation of the second term in Equation 4.3 went to zero faster than $1/B_n$. More specifically, according to Equation 4.5, any assumption which ensures $\lim_{k\to\infty} \mathscr{E}(y_k' - y_k)^2 = 0$ will suffice. One alternative is that the sequence $g_1, g_2, \cdots$ possesses a common concave modulus of continuity on J (finite or not as the case may be). That is to say, there is a function φ with the properties

$$\lim_{t\to 0} \varphi(t) = 0, \qquad 0 \leq \varphi(t_1) \leq \varphi(t_2), \qquad \frac{\varphi(t_1) + \varphi(t_2)}{2} \leq \varphi\left(\frac{t_1 + t_2}{2}\right),$$
$$(\text{all } t_2 \geq t_1 \geq 0),$$

such that

$$\sup_{\substack{|x_1 - x_2| \le t \\ (x_1, x_2) \in J}} |g_n(x_1) - g_n(x_2)| \le \varphi(t) \tag{4.19}$$

for all n. Indeed, in place of Equation 4.6, we would then have

$$\begin{aligned} |y_k' - y_k| &\le a'|g_k(u_k) - g_k(t_k)| \le a'[\varphi(|u_k - \theta|) + \varphi(|t_k - \theta|)] \\ &\le 2a'\varphi(|t_k - \theta|). \end{aligned}$$

By Equation 4.7, φ is necessarily bounded; thus, for the same reasons, $t_n \xrightarrow{P} \theta$ implies that $\mathscr{E}\varphi^2(|t_k - \theta|) \to 0$ as $k \to \infty$.

The two assumptions are clearly unrelated: the existence of (any kind of) a common modulus of continuity is an interval property of all the functions and says nothing about convergence of the sequence (continuous or otherwise) at a particular point. We have selected the continuous convergence assumption because it is implicit in Condition 1 of Theorem 4.1. This becomes clear when we turn to the question of satisfying the two provisos for some particular gain sequences.

The three gains listed below in Theorem 4.2 were introduced in Equations 2.26, 2.30, and 2.28, respectively. We have added a multiplicative gain constant in each case. The first and third gains, it will be noted, are deterministic, while the second is random. The value of θ_0 appearing in the third will usually be fixed on the basis of prior knowledge concerning the value of θ and might as well be identified with the initial guess t_1.

4.4 Large-Sample Variances for Particular Gains

THEOREM 4.2

Suppose that $\{Y_n : n = 1, 2, \cdots\}$ satisfies Assumptions A1′, A2, A3, A4, and A5‴ with $J = (-\infty, \infty)$. For all x in J, set

$$g_n(x) = \frac{|\dot{F}_n(x)|}{b_n}, \qquad c = \limsup_n \frac{d_n}{b_n}.$$

Assume that the functions $g_1, g_2, \cdots$ are continuously convergent at θ to a number γ. Furthermore, suppose that

$$\operatorname{Var} Y_n = \sigma^2 \qquad \text{and} \qquad \sup_n \mathscr{E}|Y_n - F_n(\theta)|^{2+\delta} < \infty$$

for some $\delta > 0$. Let t_1 be arbitrary, and let

$$t_{n+1} = t_n + a_n(t_1, \cdots, t_n)[Y_n - F_n(t_n)] \qquad (n = 1, 2, \cdots),$$

where $\{a_n\}$ is one of the following gains:

Gain	$a_n(x_1, \cdots, x_n)$	
1	$A_1 \operatorname{sgn} \dot{F}_n \dfrac{b_n}{B_n^2}$	
2	$A_2 \dfrac{\dot{F}_n(x_n)}{\sum_{k=1}^{n} \dot{F}_k^2(x_k)}$	
3	$A_3 \dfrac{\dot{F}_n(\theta_0)}{\sum_{k=1}^{n} \dot{F}_k^2(\theta_0)}$	$(\theta_0 \in J)$.

The A's are suitably chosen constants. Define the function

$$V(x) = \frac{x^2}{2x - 1} \qquad (\tfrac{1}{2} < x < \infty).$$

Then, as $n \to \infty$,

$$\sqrt{\sum_{k=1}^{n} \dot{F}_k^2(\theta)}\, (t_n - \theta)/\sigma \sim N(0, Q^2),$$

with corresponding variances given by

Gain	Q^2	
1	$V(A_1\gamma)$,	provided $A_1 > \frac{1}{2}$,
2	$V(A_2)$,	provided $A_2 > c^2/2$,
3	$V\left(A_3 \dfrac{\gamma}{\gamma_0}\right)$,	provided $A_3 > c^2/2$, and $g_n(\theta_0) \to \gamma_0$ as $n \to \infty$.

In every case, the same limiting distribution obtains if, in the norming sequence, $\dot{F}_k(\theta)$ is replaced by $\dot{F}_k(t_k)$ for $k = 1, 2, \cdots, n$.

Proof. To save space we set

$$b_{nk} = \frac{b_k^2}{B_n^2} \qquad (k = 1, 2, \cdots, n).$$

This constitutes a normalized Toeplitz matrix because each column tends to zero as $n \to \infty$ and the row sums are identically one. Thus,

$$\lim_n f_n = f \quad \text{implies} \quad \lim_n \sum_{k=1}^{n} b_{nk} f_k = f \tag{4.20}$$

by the Toeplitz Lemma (Knopp, 1947, p. 75). This fact will be used repeatedly.

To apply Theorem 4.1, we must first verify that the number L, defined in its hypothesis, exceeds $\frac{1}{2}$ for each of the three gains under consideration. For Gain 1 this is immediate because L is A_1, and the latter is presumed chosen larger than $\frac{1}{2}$. For gain 2 we have (and the same clearly will be the case for Gain 3)

$$L = \liminf_n \inf_{\mathbf{x}\in J^{(n)}} \frac{B_n^2}{b_n} |a_n^{(2)}(\mathbf{x})| = \liminf_n \inf_{\mathbf{x}\in J^{(n)}} A_2 \frac{g_n(x_n)}{\sum_{k=1}^n b_{nk} g_k^2(x_k)}$$

$$\geq \frac{A_2}{\limsup_n \sum_{k=1}^n b_{nk}\left(\frac{d_k}{b_k}\right)^2} \geq \frac{A_2}{c^2} > \frac{1}{2}.$$

In the last line we have used the fact that

$$\limsup_n \sum_{k=1}^n b_{nk} f_k \leq \limsup_n f_n \tag{4.20a}$$

if $0 < \inf_n f_n \leq \sup_n f_n < K < \infty$. Indeed, if we set $\bar{f} = \limsup_n f_n$, there corresponds, to any $\varepsilon > 0$, a finite index n_0 such that $f_n < \bar{f} + \varepsilon$ for all $n > n_0$. For such indices, we have

$$y_n = \sum_{k=1}^n b_{nk} f_k < K \sum_{k=1}^{n_0} b_{nk} + \bar{f} + \varepsilon.$$

The first term tends to zero as $n \to \infty$; hence, there is an $n_1 > n_0$ such that it remains less than ε for all $n > n_1$. Thus, for all sufficiently large n,

$$y_n < \bar{f} + 2\varepsilon,$$

from which the asserted conclusion follows because ε was arbitrary.

The problem is to prove that Conditions 1 and 2 in the statement of Theorem 4.1 are satisfied with values of (λ, μ) which yield the asserted formulas for Q^2. To do this we set

$$\begin{aligned} S_n^2 &= \frac{1}{B_n^2} \sum_{k=1}^n \dot{F}_k^2(t_k) = \sum_{k=1}^n b_{nk} g_k^2(t_k), \\ \Sigma_n^2(x) &= \frac{1}{B_n^2} \sum_{k=1}^n \dot{F}_k^2(x) = \sum_{k=1}^n b_{nk} g_k^2(x), \end{aligned} \tag{4.21}$$

for x in J. Let y_n and z_n have the meanings respectively given in Equations 4.1 and 4.11 as functions of $t_1, \cdots, t_n$:

$$y_n = \frac{B_n^2}{b_n} |a_n| \, g_n(t_n) \qquad \text{and} \qquad z_n = \frac{B_n^4}{b_n^2} |a_n|^2.$$

The first two columns of the following table are proportional to these sequences for the listed gains.

Gain	y_n/A	z_n/A^2	λ/A	μ/A^2
1	$g_n(t_n)$	1	γ	1
2	$\dfrac{g_n^{\,2}(t_n)}{S_n^{\,2}}$	$\dfrac{g_n^{\,2}(t_n)}{S_n^{\,4}}$	1	$\dfrac{1}{\gamma^2}$
3	$\dfrac{g_n(\theta_0)g_n(t_n)}{\Sigma_n^{\,2}(\theta_0)}$	$\dfrac{g_n^{\,2}(\theta_0)}{\Sigma_n^{\,4}(\theta_0)}$	$\dfrac{\gamma}{\gamma_0}$	$\dfrac{1}{\gamma_0^{\,2}}$

(4.22)

We now show, in each of the three cases, that

$$\mathscr{E}(y_n - \lambda)^2 = o(1), \tag{4.23a}$$

$$\mathscr{E}(z_n - \mu)^2 = o(1), \tag{4.23b}$$

as $n \to \infty$ for the corresponding (λ, μ) given in the third and fourth columns.

First of all, however, we note that each of the asserted λ values exceeds $\frac{1}{2}$. Indeed, since

$$1 \leq \liminf_n g_n(x) \leq \limsup_n g_n(x) \leq c$$

for all $x \in J$, any limiting values of $g_n(x)$ must belong to the interval $[1, c]$; in particular, γ and γ_0. Thus, in the case of Gain 3, $\lambda = A_3(\gamma/\gamma_0) > (c^2/2)(1/c) = c/2 \geq \frac{1}{2}$ with equality only when $c = 1$, in which case we say the problem is *asymptotically linear*.

With regard to Gain 1: The hypothesized continuous convergence of the g_n's at θ to γ immediately allows us to infer $g_n(t_n) \xrightarrow{P} \gamma$ from $t_n \xrightarrow{P} \theta$. But the g_n's are bounded, so $g_n(t_n) \to \gamma$ in mean square.

With regard to Gain 2: We consider the identity

$$\frac{y_n}{A_2} - 1 = \gamma^2 \left[\frac{1}{S_n^{\,2}} - \frac{1}{\Sigma_n^{\,2}(\theta)}\right] + \left[\frac{\gamma^2}{\Sigma_n^{\,2}(\theta)} - 1\right] + \frac{1}{S_n^{\,2}}\,[g_n^{\,2}(t_n) - \gamma^2].$$

According to Equations 4.7 and 4.21,

$$1 \leq S_n^{\,2}, \Sigma_n^{\,2}(x) \leq c_0^{\,2} < \infty$$

for all x in J; hence

$$\left|\frac{y_n}{A_2} - 1\right| \leq \gamma^2|S_n^{\,2} - \Sigma_n^{\,2}(\theta)| + |\Sigma_n^{\,2}(\theta) - \gamma^2| + |g_n^{\,2}(t_n) - \gamma^2|. \tag{4.24}$$

By the same argument used for Gain 1, the third term goes to zero in mean square. For the first, from Equation 4.21,

$$|S_n^2 - \Sigma_n^2(\theta)| \le \sum_{k=1}^{n} b_{nk}[|g_k^2(t_k) - \gamma^2| + |g_k^2(\theta) - \gamma^2|].$$

According to Theorem 2.3, Condition 3, $t_k \xrightarrow{\text{a.s.}} \theta$ as $k \to \infty$. Thus, and again by the continuous convergence assumption, the random variable enclosed in square brackets tends a.s. to zero as $k \to \infty$. Now the implication of Equation 4.20 is valid when the f_n's and f are random variables and "lim" is replaced by "a.s. lim" (but is not, incidentally, when replaced by "P lim"). Hence, $S_n^2 - \Sigma_n^2(\theta) \xrightarrow{\text{a.s.}} 0$ and, because the variates are bounded, we have

$$\lim_n \mathscr{E}|S_n^2 - \Sigma_n^2(\theta)|^2 = 0. \tag{4.25}$$

Furthermore,

$$\lim_n \Sigma_n^2(\theta) = \gamma^2. \tag{4.26}$$

Equation 4.23*a* therefore follows after we square Equation 4.24 and take expectations.

To establish Equation 4.23*b*, we note that z_n/A_2^2 is $1/S_n^2$ times y_n/A_2. Thus,

$$\frac{z_n}{A_2^2} - \frac{1}{\gamma^2} = \frac{1}{S_n^2}\left(\frac{y_n}{A_2} - 1\right) + \frac{1}{S_n^2} - \frac{1}{\gamma^2},$$

so that, since γ is also no smaller than unity,

$$\left|\frac{z_n}{A_2^2} - \frac{1}{\gamma^2}\right| \le \left|\frac{y_n}{A_2} - 1\right| + |S_n^2 - \gamma^2|.$$

It follows from the results of the previous paragraph that this bound goes to zero in mean square as $n \to \infty$.

With regard to Gain 3: If we use the additional assumption that $\{g_n\}$ is convergent at the selected point θ_0, the same type of argument used in the preceding paragraphs establishes Equations 4.23*a* and 4.23*b* for the asserted λ and μ in Equation 4.22.

We have thus verified all the (unassumed) hypotheses of Theorem 4.1. In view of Equations 4.25 and 4.26, we have

$$S_n = \frac{1}{B_n}\sqrt{\sum_{k=1}^{n} \dot{F}_k^2(t_k)} \xrightarrow{P} \gamma$$

as $n \to \infty$, and the limit is a sure one when every t_k is replaced by θ. Hence, by the conclusion of Theorem 4.1, we have

$$\sqrt{\sum_{k=1}^{n} \dot{F}_k^2(t_k)}\,(t_n - \theta)/\sigma \sim \gamma B_n(t_n - \theta)/\sigma \sim N\left(0, \frac{\gamma^2\mu}{2\lambda - 1}\right).$$

It remains to note that Q^2 is precisely $\gamma^2\mu/(2\lambda - 1)$ when we substitute the values of λ and μ given in Equation 4.22. Q.E.D.

The following result clearly requires no independent proof.

THEOREM 4.2′

Let the hypotheses of Theorem 4.2 hold over an interval $J = [\xi_1, \xi_2]$, with at least one of the end points finite, where the interval is so chosen that θ is an interior point. In addition, assume there is a finite integer $p \geq 2$ such that

$$B_n^{p-2}b_n \to \infty$$

with n. Also suppose that

$$\sup_n \mathscr{E}[Y_n - F_n(\theta)]^{2p} < \infty.$$

If t_1 is arbitrary and

$$t_{n+1} = [t_n + a_n(t_1, \cdots, t_n)[Y_n - F_n(t_n)]]_{\xi_1}^{\xi_2} \qquad (n = 1, 2, \cdots),$$

where $\{a_n\}$ is one of the three gains listed in Theorem 4.2, then the conclusion of Theorem 4.2, under its provisos, holds for these truncated estimates.

4.5 Other Gains

After examining the proof of Theorem 4.2 it is clear how to deduce the asymptotic distribution of the estimates generated by any restricted gain sequence which can be appropriately expressed in terms of the g_n's and, furthermore, we know what additional conditions (if any) should be imposed in order to do so. Consider, for example,

$$a_n = A\frac{b_n^2}{B_n^2}\frac{1}{\dot{F}_n(\xi_n)} = \operatorname{sgn}\dot{F}_n A\frac{b_n}{B_n^2}\frac{1}{g_n(\xi_n)}, \tag{4.27}$$

where $\{\xi_n\}$ is any sequence of random variables, taking values in J, which converges in probability to a limit ξ as $n \to \infty$. We first restrict A to ensure $L > \frac{1}{2}$; in this case,

$$A > \frac{c}{2}.$$

Mean-square limits of the random variables y_n and z_n must next be guaranteed. Here we have

$$y_n = A\frac{g_n(t_n)}{g_n(\xi_n)}, \quad z_n = \frac{A^2}{g_n^2(\xi_n)};$$

and therefore we require that $g_1, g_2, \cdots$ be continuously convergent at the point ξ to, say, γ_ξ. The mean-square limits are then

$$\lambda = A\frac{\gamma}{\gamma_\xi} > \frac{1}{2}, \qquad \mu = \frac{A^2}{\gamma_\xi^2},$$

after using weak convergence and boundedness. The variance is thus

$$Q^2 = \frac{\gamma^2\mu}{2\lambda - 1} = V\left(A\frac{\gamma}{\gamma_\xi}\right). \tag{4.28}$$

If ξ_n is defined as the value of x at which the infimum of $|\dot{F}_n(x)|$ is assumed, that is, $\dot{F}_n(\xi_n) = b_n/\text{sgn}\,\dot{F}_n$, then $\gamma_\xi = 1$, and Equation 4.27 becomes Gain 1. If we take $\xi_n = t_n$, then $\gamma_\xi = \gamma$ and the variance, Equation 4.28, is the same as that for Gain 2, although the gains are algebraically different. Finally, the same is true for $\xi_n = \theta_0$ and Gain 3. The fact that both Gain 2 and Gain 3 are easier to compute than Equation 4.27 is reflected in the stronger limitation $A > c^2/2$.

4.6 Gain Comparison and Choice of Gain Constants

We should compare the estimates in Theorem 4.2 on the basis not only of their relative asymptotic efficiencies but also the amount of labor involved in calculating and using the corresponding gain sequences. We have numbered the gains independently of any such considerations, but in the order of the increasing analytical restrictions imposed in the Q^2-table.

It is clear that there is no universal ordering of the costs C_1, C_2, and C_3 associated with using the respective gains (if, indeed, such a numerical value can even be assigned), and that the problem must be treated in the light of the particular application. However, some rather vague general relationships can be cited. Thus, Gains 1 have the advantage over Gains 2 in that they can be computed before the data acquisition and thereby decrease computation time. Such is also true of Gain 3, which has the added advantage that it does not require locating the derivative minima but, rather, just their calculation at the selected initial guess. Thus, we might write $C_3 < C_1 < C_2$. But unless there is a recursive relation between successive b_n's, we are faced with the problem of storing the entire Gain 1 sequence. Gain 2, on the other hand, can be inversely generated on line by means of the recursion

$$\frac{1}{a_{n+1}} = \frac{\dot{F}_n(t_n)}{\dot{F}_{n+1}(t_{n+1})}\frac{1}{a_n} + \frac{\dot{F}_{n+1}(t_{n+1})}{A_2} \qquad (n = 1, 2, \cdots),$$

which is to be initialized by $1/a_1 = \dot{F}_1(t_1)/A_2$. The Gain 3 sequence is computed in the same fashion with t_n replaced by θ_0 for all $n \geq 1$, and

A_2 by A_3. Thus, storage considerations suggest inverting the order to $C_3 < C_2 < C_1$. Still, there are problems in which the minimum value of $|\dot{F}_n(x)|$ is taken on at one and the same end point of J for every n. (Such is the case for the example worked out at the end of Chapter 5.) This leads to a further change: $C_1 = C_3 < C_2$.

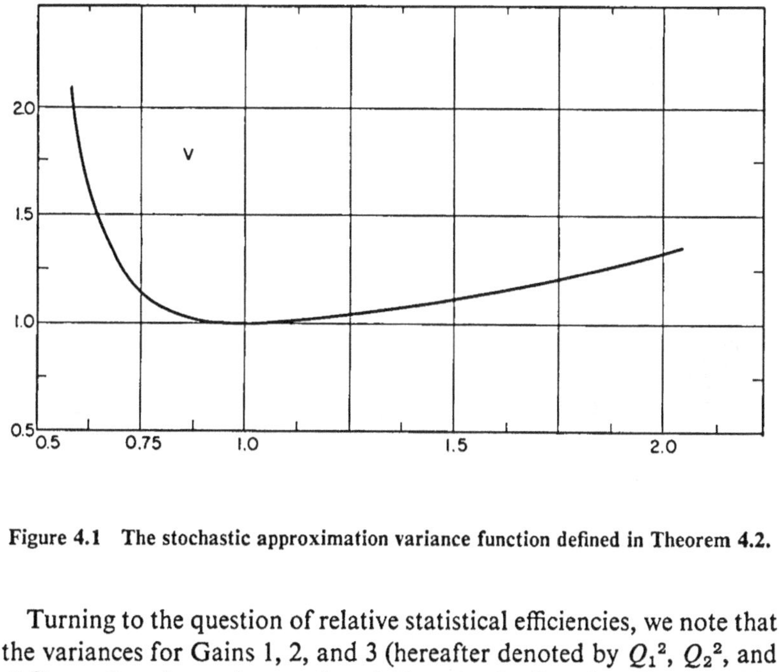

Figure 4.1 **The stochastic approximation variance function defined in Theorem 4.2.**

Turning to the question of relative statistical efficiencies, we note that the variances for Gains 1, 2, and 3 (hereafter denoted by Q_1^2, Q_2^2, and Q_3^2) are functions of several variables via the function V, plotted in Figure 4.1:

$$Q_1^2 = Q_1^2(A_1, \gamma), \qquad Q_2^2 = Q_2^2(A_2, c), \qquad Q_3^2 = Q_3^2(A_3, c, \gamma_0, \gamma).$$

For Gain 1, A_1 must be chosen in the open interval $(\frac{1}{2}, \infty)$. For Gain 2, A_2 must be chosen in $(c^2/2, \infty)$, while for Gain 3, θ_0 can be chosen by the experimenter (this determines γ_0), and then A_3 must be chosen in $(c^2/2, \infty)$. For any particular choices of the A_i, it is not hard to exhibit regressions such that each gain is, in turn, "optimal" (has minimal Q^2) for some value of the parameter θ. Thus, the question of "which gain to use" has no quick answer.

As a possible guideline for comparing the three types of gains for a particular regression when θ (hence γ) is not known, we might adopt a "minimax" criterion for choosing the A_i and then compare the variance functions as γ varies over its domain. That is to say, we compare

$$Q_1^2(A_1^*, \gamma), \qquad Q_2^2(A_2^*, c), \qquad \text{and} \qquad Q_3^2(A_3^*, c, \gamma_0, \gamma),$$

as γ varies, where the A_i^* are chosen from their respective constraint sets to minimize

$$\max_{\theta} Q_1^2(A_1, \gamma), \qquad Q_2^2(A_2, c), \qquad \text{and} \qquad \max_{\theta} Q_3^2(A_3, c, \gamma_0, \gamma),$$

respectively. (For any particular regression problem c is known, and we will assume that θ_0, hence γ_0, has been determined by considerations of "nominal" parameter values.) As we will see, only the first function achieves its minimum on the constraint set.

We "minimize" Q_2^2 by

$$A_2^* = \begin{cases} 1 & \text{for } 1 \le c < \sqrt{2}, \\ \dfrac{c^2}{2} + 0 & \text{for } c \ge \sqrt{2}, \end{cases} \tag{4.29}$$

where the $+0$ indicates the lack of a minimum over $A_2 > \frac{1}{2}$. With regard to Gains 1 and 3, maximization over $\theta \in J$ is the same as maximization over all real numbers $\gamma \in [1, c]$. For the former we see that

$$\max_{1 \le \gamma \le c} V(A\gamma) = \max\{V(A), V(Ac)\} \qquad (A > \tfrac{1}{2}). \tag{4.30}$$

This is minimized by the value of $A = A(c)$ which makes $V(A)$ and $V(Ac)$ equal. The solution is simply

$$A_1^* = \frac{1}{2}\left(1 + \frac{1}{c}\right) \qquad (1 \le c < \infty), \tag{4.31}$$

which decreases monotonically from 1 to $\frac{1}{2}$. The situation for Gain 3 is a bit more complicated. Now we seek to minimize Equation 4.30 with respect to A, subject to the restriction $A > c^2/2\gamma_0 \ge \frac{1}{2}$ (A is A_3/γ_0). The solution is given by the right-hand side of Equation 4.31, provided that $\gamma_0 > c^3/(c + 1)$, which cannot take place unless c is small enough so that $c^3 < c(c + 1)$. This leads to a rather complicated formula for Gain 3:

$$A_3^* = \begin{cases} \dfrac{\gamma_0}{2}\left(1 + \dfrac{1}{c}\right) & \text{if } c^3 < c(c+1) \text{ and } \gamma_0 > \dfrac{c^3}{c+1}, \\ \dfrac{c^2}{2} + 0 & \text{otherwise.} \end{cases} \tag{4.32}$$

The values of the variance resulting from the choices of Equations 4.29, 4.31, and 4.32 are

$$Q_1^2 = V\left(\frac{c+1}{2c}\gamma\right),$$

$$Q_2^2 = \begin{cases} 1 & \text{if } 1 \le c < \sqrt{2}, \\ V\left(\frac{c^2}{2}\right) + 0 & \text{if } c \ge \sqrt{2}, \end{cases} \tag{4.33}$$

$$Q_3^2 = \begin{cases} V\left(\frac{c+1}{2c}\gamma\right) & \text{if } c^3 < c(c+1) \text{ and } \gamma_0 > \frac{c^3}{c+1}, \\ V\left(\frac{c^2}{2}\frac{\gamma}{\gamma_0}\right) + 0 & \text{otherwise,} \end{cases}$$

where $1 \le \gamma, \gamma_0 \le c$. We see that every $Q_i^2 \ge 1$ with equality when and only when $c = 1$.

The same is true for the simpler choices

$$A_1 = 1 \quad \text{and} \quad A_2 = A_3 = c^2, \tag{4.34}$$

which meet the provisos in all problems. The corresponding variances are

$$Q_1^2 = V(\gamma), \qquad Q_2^2 = V(c^2), \qquad Q_3^2 = V\left(c^2\frac{\gamma}{\gamma_0}\right). \tag{4.35}$$

It is interesting to note in Equation 4.33 that $Q_3^2 < Q_2^2$ whenever $c \ge \sqrt{2}$ and

$$\gamma_0 = \lim_n g_n(\theta_0) > \text{a.s.} \lim_n g_n(t_n) = \gamma.$$

The same is true in Equation 4.35 for every c. Thus, a fortuitous choice for θ_0 will make the estimates based on the more easily computed Gain 3 asymptotically more efficient than those based on Gain 2.

In the next chapter we limit our consideration to sequences $g_1, g_2, \cdots$ that converge uniformly on J to a continuous limit g. We then, at an increased computational cost, iterate in a certain transformed parameter space defined only by g and invert back to J at each step. The result, as might already be anticipated, is that $Q^2 = V(1) = 1$ for all three gains, because the transformation will be selected to force asymptotic linearity, that is, $c = 1$.

Before doing this, however, we point out that our methods of proof in this chapter (as was the case in Chapter 3) are readily adapted, after appropriate reinterpretation of the symbols, to yield asymptotic normality for a general class of stochastic approximation processes.

4.7 A General Stochastic Approximation Theorem

THEOREM 4.3

For every real number x and positive integer n, let $Z_n(x)$ be an observable random variable. Corresponding to a given sequence of constants $\alpha_1, \alpha_2, \cdots$, recursively define

$$t_{n+1} = t_n - \alpha_n z_n \qquad (n = 1, 2, \cdots; t_1 \text{ arbitrary}),$$

where z_n is a random variable whose conditional distribution, given $t_1 = \xi_1, t_2 = \xi_2, \cdots, t_n = \xi_n$, is the same as the distribution of $Z_n(\xi_n)$. Let

$$G_n(x) = \mathscr{E} Z_n(x)$$

have a zero which converges to a finite number θ as $n \to \infty$:

$$G_n(\theta_n) = 0, \qquad \lim_n \theta_n = \theta.$$

Furthermore, suppose that $(x - \theta_n)G_n(x)$ is one-signed (say positive) for all $x \neq \theta_n$. We impose the following further conditions.

1. There exists $b_n > 0$ and a number γ such that

$$g_n(x) = \begin{cases} \dfrac{G_n(x)}{b_n(x - \theta_n)} & \text{if } x \neq \theta_n, \\ \gamma & \text{if } x = \theta_n \end{cases}$$

satisfies $1 \leq g_n(x) \leq c_0 < \infty$ for all x and all $n > n_0$. (We can always redefine b_n so that any strictly positive value of $\inf_{n > n_0, x} g_n(x)$ is unity.)
2. $g_n(x)$ is continuously convergent at θ to γ.
3. $B_n^2 = b_1^2 + \cdots + b_n^2 \to \infty$ with n and $\Sigma b_n^4/B_n^4 < \infty$.
4. $\theta_n = \theta + o(1/B_n)$.
5. $\sup_{n,x} \mathscr{E}|Z_n(x) - G_n(x)|^{2+\delta} < \infty$ for some $\delta > 0$ and $\operatorname{Var} Z_n(x)$ is continuously convergent at θ to a number σ^2.

Then, if $\{\alpha_n\}$ is any positive number sequence so chosen that

$$\gamma \lim_n \frac{B_n^2}{b_n} \alpha_n = \lambda > \tfrac{1}{2},$$

the random variables

$$\frac{\gamma B_n(t_n - \theta)}{\sigma} \qquad (n = 1, 2, \cdots)$$

have a large-sample normal distribution about 0 with variance

$$V(\lambda) = \frac{\lambda^2}{2\lambda - 1}.$$

Toward Proof. In the case $\theta_n = \theta$ for all n (that is, a common root of the mean-value functions) and $\operatorname{Var} Z_n(x) = \sigma^2$ for all n and x, the validity of the assertion is an almost immediate consequence of Theorem 4.1, after we identify $g_n(x)$ as defined here with $|\dot{F}_n(u_n)|/b_n$, etc. (Compare with Equations 3.14 through 3.17.) But an examination of the proof of Theorem 4.1 shows that the perturbation of any quantities which take values in the parameter space by terms which go to zero faster than $1/B_n$ has no effect on the limiting behavior of $\{t_n\}$. Moreover, continuous convergence of bounded functions of such quantities at θ yields the same conclusion that results from assuming the limiting values to start with.

The conclusion of Theorem 4.3 is precisely the conclusion of Burkholder's (1956) Theorem 3 when

$$b_n \sim \frac{1}{n^{1/2-\xi}} \qquad (0 < \xi \leq \tfrac{1}{2}) \tag{4.36}$$

as $n \to \infty$. (This is not obvious until the symbols in the two statements are properly related.) As already noted at the end of Chapter 3, our Condition 3 is much less restrictive than Equation 4.36. Furthermore, Burkholder assumes that all moments of $Z_n(x) - G_n(x)$ are finite, albeit only throughout some neighborhood of θ. Condition 5, at least from the point of view of application, is in most instances weaker. Indeed, the distribution of the "noise", $Z_n(x) - G_n(x)$, usually depends on x in a rather trivial fashion and is often independent of the adjustable parameter. On the other hand, high-order absolute moments are infinite in some problems. Finally, Burkholder's assumption that

$$\sup_{n,x} \frac{|G_n(x)|}{1 + |x|} < \infty, \qquad \Sigma\, \alpha_n \inf_{|x-\theta|>\delta} |G_n(x)| = \infty$$

for every $\delta > 0$ is weaker than our Condition 1, provided that $\limsup_n b_n < \infty$. But, as already pointed out, we make no such limitation as Equation 4.36.

5. Asymptotic Efficiency

The third gain considered in Theorem 4.2,

$$a_n^{(3)} = \frac{\dot{F}_n(\theta_0)}{\sum_{k=1}^{n} \dot{F}_k^2(\theta_0)},$$

is appropriate in many applications. As we have noted, it is computationally cheaper than the "adaptive" second gain, and it can lead to estimates that are more efficient in large samples. However, the existence of a stable limiting distribution for these estimates should not depend on the value of our initial guess, $t_1 = \theta_0$. Hence, the Gain-3 proviso (that $\{g_n(x)\}$ converge at the particular point $x = \theta_0$) ought to be replaced by the assumption that the sequence possess a limit, say $g(x)$, at every x in J. If, in addition, we require that this convergence be uniform on J and that the limit function be continuous, there will be continuous convergence at every point of J (in particular, at θ, as also required in Theorem 4.2). Indeed, if ξ is arbitrary in J and $\{x_n\}$ is any sequence tending to ξ, then

$$|g_n(x_n) - g(\xi)| \le \sup_{x \in J} |g_n(x) - g(x)| + |g(x_n) - g(\xi)|$$

tends to zero as $n \to \infty$. Therefore, throughout Chapter 4 we might just as well have hypothesized uniform convergence on J of $g_1, g_2, \cdots$ to a continuous limit function g. The latter assumption will be a consequence of the former when J is bounded, provided that each member of the sequence is continuous. In the conclusion of Theorem 4.2, we now identify γ and γ_0 with g evaluated at θ and θ_0, respectively.

5.1 Asymptotic Linearity

Having thus hypothesized the existence and knowledge of a limit function, we can now construct recursive estimation schemes that, as will subsequently be shown, become asymptotically efficient when (and only when) the observational errors are Gaussian. We demonstrate this for the case of a bounded interval J, which is henceforth identified with the prior knowledge space of the parameter. From the applied point of view, this does not constitute a significant restriction.

By way of introduction, we note that our new assumptions allow us to write

$$F_n(x) \cong \operatorname{sgn} \dot{F}_n b_n \int_{\xi_1}^{x} g(\xi)\, d\xi + K_n \tag{5.1}$$

as $n \to \infty$ for all x in J (because the interchange of the integration and limit operations is permissible). The regression functions are therefore *asymptotically linear in the values of the integral.* It is reasonable, therefore, to estimate recursively the parameter value

$$\varphi = \int_{\xi_1}^{\theta} g(\xi)\, d\xi$$

and invert back at each step to obtain the θ-estimate. This is, in fact, the method analyzed in the following theorem.

In some rather simple problems, Equation 5.1 is an equality for every n (and the major portion of the proof of Theorem 5.1 is obviated). For instance, if $F_n(\theta) = k_n\theta^3$, and J is any finite interval that does not include the origin, then $g_n(x) = (x/\xi_1)^2$ for all n. In such a situation, we would estimate θ^3 by linear least squares and then take the cube root.

5.2 Increased Efficiency via Transformation of the Parameter Space

THEOREM 5.1

Let Assumptions A1′, A2, A3, A4, and A5‴ hold, where $J = [\xi_1, \xi_2]$ is any finite interval containing θ as an interior point. For $n \geq 1$, let

$$g_n(x) = \frac{|\dot{F}_n(x)|}{b_n}$$

be continuous at every x in J, and suppose that

$$\sup_{x \in J} |g_n(x) - g(x)| \to 0$$

as $n \to \infty$. In addition, suppose that

$$\text{Var } Y_n = \sigma^2 \text{ and } \sup_n \mathscr{E}[Y_n - F_n(\theta)]^{2p} < \infty,$$

where $p \geq 2$ is the smallest integer (assumed to exist) for which

$$\lim_n B_n^{p-2} b_n = +\infty.$$

For x in J, define

$$\Phi(x) = \int_{\xi_1}^{x} g(\xi)\, d\xi,$$

which takes values in $J^* = [0, \Phi(\xi_2)]$, and let $\Psi = \Phi^{-1}$ be the inverse function (which exists because g is positive and bounded). For y in J^*, define

$$F_n^*(y) = F_n(\Psi(y)),$$

$$b_n^* = \inf_{y \in J^*} |\dot{F}_n^*(y)|, \qquad B_n^{*2} = \sum_{k=1}^{n} b_k^{*2},$$

where the dot means differentiation with respect to y. Let t_1^* be arbitrary, and

$$t_{n+1}^* = [t_n^* + a_n^*[Y_n - F_n^*(t_n^*)]]_0^{\Phi(\xi_2)} \qquad (n = 1, 2, \cdots),$$

where a_n^* is any one of the quantities

$$\operatorname{sgn} \dot{F}_n^* \frac{b_n^*}{B_n^{*2}}, \qquad \frac{\dot{F}_n^*(t_n^*)}{\sum_{k=1}^{n} \dot{F}_k^{*2}(t_k^*)}, \qquad \frac{\dot{F}_n^*(\varphi_0)}{\sum_{k=1}^{n} \dot{F}_k^{*2}(\varphi_0)},$$

and φ_0 is an arbitrarily selected point in J^*. Finally, let

$$t_n = \Psi(t_n^*) \qquad (n = 1, 2, \cdots).$$

Then, as $n \to \infty$, we have

$$\sqrt{\sum_{k=1}^{n} \dot{F}_k^2(\theta)}\,(t_n - \theta)/\sigma \sim N(0, 1),$$

and the same holds true when every $\dot{F}_k(\theta)$ is replaced by $\dot{F}_k(t_k)$ in the norming sequence.

Proof. Letting $\varphi = \Phi(\theta)$ denote the unknown parameter in the transformed space, we have, by definition, $\mathscr{E} Y_n = F_n(\theta) = F_n^*(\varphi)$. The proof of the theorem falls into two parts. We first show that the starred problem is asymptotically a linear one. Since each of the starred gains has a gain constant purposely chosen to be unity, the Q^2-table of Theorem 4.2 with $c = 1$ will then give (via Theorem 4.2′)

$$\sqrt{\sum_{k=1}^{n} \dot{F}_k^{*2}(\varphi)}\,(t_n^* - \varphi)/\sigma \sim N(0, 1) \tag{5.2}$$

in all three cases. The second part of the proof will yield the desired conclusion by the "delta method."

The initial step, then, is to show that our assumption that $\{F_n\}$ obeys Assumptions A1′ through A5‴ on J implies that $\{F_n^*\}$ does on J^*. The basic relation for doing this is

$$\begin{aligned}\dot{F}_n^*(y) &= \dot{F}_n(\Psi(y))\frac{d\Psi(y)}{dy} = \frac{\dot{F}_n(\Psi(y))}{\left.\dfrac{d\Phi(x)}{dx}\right|_{x=\Psi(y)}}\\ &= \operatorname{sgn}_J \dot{F}_n b_n \frac{g_n(\Psi(y))}{g(\Psi(y))}.\end{aligned} \tag{5.3}$$

We immediately see that the sign of $\dot{F}_n^*(y)$ over $y \in J^*$ is constant and the same as that of $\dot{F}_n(x)$ over $x \in J$. Furthermore, we have

$$b_n^* = b_n \inf_J \frac{g_n(x)}{g(x)}, \qquad d_n^* = b_n \sup_J \frac{g_n(x)}{g(x)}.$$

Since the range of the limit function cannot be larger than that shared by every member of the sequence, Equation 4.7 yields

$$b_n^* \geq \frac{b_n}{c_0}, \qquad d_n^* \leq b_n c_0 \qquad (n = 1, 2, \cdots).$$

Thus, not only are Assumptions A1′ through A5‴ satisfied by the starred infimums and supremums, but also

$$\lim_n B_n^{*p-2} b_n^* = +\infty$$

for the same hypothesized integer p.

We use the uniform convergence to show that

$$c^* = \limsup_n \frac{d_n^*}{b_n^*} = 1.$$

The ratio

$$\frac{d_n^*}{b_n^*} = \frac{\sup_J [g_n(x)/g(x)]}{\inf_J [g_n(x)/g(x)]}$$

actually converges because both numerator and denominator tend to unity. Indeed, we have

$$\left|\sup_J \frac{g_n(x)}{g(x)} - 1\right| \le \sup_J \left|\frac{g_n(x)}{g(x)} - 1\right| \le \sup_J |g_n(x) - g(x)| = o(1)$$

and, similarly, $\inf_J g_n(x)/g(x) \to 1$ as $n \to \infty$. It is *a fortiori* true that

$$g_n^*(y) = \frac{|\dot{F}_n^*(y)|}{b_n^*} \to 1 = g^*(y)$$

(uniformly) on J^*. In other words, in the Q^2-table of Theorem 4.2, we are to read $c = \gamma = \gamma_0 = 1$. This establishes Equation 5.2.

To obtain the limiting distribution of the estimates t_n of $\theta = \Psi(\varphi)$, we expand in the usual fashion:

$$\Psi(t_n^*) = \Psi(\varphi) + \dot{\Psi}(\varphi)(t_n^* - \varphi) + [\dot{\Psi}(v_n) - \dot{\Psi}(\varphi)](t_n^* - \varphi),$$

where v_n is some random point such that $|v_n - \varphi| < |t_n^* - \varphi|$. The derivative

$$\dot{\Psi}(y) = \frac{1}{g(\Psi(y))}$$

is continuous and nonzero at every y in J^*. From Equation 5.2, we see that $t_n^* \xrightarrow{P} \varphi$; hence $\dot{\Psi}(v_n) \xrightarrow{P} \dot{\Psi}(\varphi)$. Thus, after we multiply through by the appropriate norming sequence and use Equation 5.2 as written, it follows that

$$\sqrt{\sum_{k=1}^{n} \dot{F}_k^{*2}(\varphi)}\,(t_n - \theta) = \dot{\Psi}(\varphi)\sqrt{\sum_{k=1}^{n} \dot{F}_k^{*2}(\varphi)}\,(t_n^* - \varphi) + o_p(1)O_p(1)$$
$$\sim N(0, \dot{\Psi}^2(\varphi)\sigma^2).$$

But according to the leading equality in Equation 5.3,

$$\dot{F}_k^*(\varphi) = \dot{F}_k(\theta)\dot{\Psi}(\varphi),$$

so that

$$\sqrt{\sum_{k=1}^{n} \dot{F}_k^2(\theta)}\,(t_n - \theta) \sim N(0, \sigma^2), \tag{5.4}$$

which is the asserted distribution.

Now, from Theorem 2.3, Condition 3, we know that $t_n^* \xrightarrow{\text{a.s.}} \varphi$ and, hence, $t_n \xrightarrow{\text{a.s.}} \theta$ by continuity of Ψ. It follows by precisely the same argument used with regard to Gain 2 in the proof of Theorem 4.2 that

$$\frac{\sum_{k=1}^{n} \dot{F}_k^2(t_k)}{\sum_{k=1}^{n} \dot{F}_k^2(\theta)} = \frac{S_n^2}{\Sigma_n^2(\theta)} \xrightarrow{\text{a.s.}} 1,$$

in the notation of Equation 4.21. This combines with Equation 5.4 to prove the final statement. Q.E.D.

The appropriately normalized deviations $t_n - \theta$ of Theorem 5.1 have a large sample $N(0, 1)$ distribution for any of the listed gains, none of which contain undetermined constants (which is as it should be). The result is true without variance dependence on unknowns or quantitative restriction on the limit function g which occurred in the conclusion of Theorem 4.2. The computation of these transformed estimates is clearly more time-consuming, but this is the price we must pay for the improvement in variance. Since all three gains yield the same limiting distribution, the computationally cheapest third one will usually be used.

5.3 Asymptotic Efficiency and Summary Theorem

The question naturally arises as to how these estimates (or, as a matter of fact, any of our estimates) compare statistically with the still more computationally expensive method of Maximum Likelihood (abbreviated ML). A good deal of the following discussion pertaining to this topic is standard material; it is included for the sake of completeness.

We are going to assume that the observational errors $W_n = Y_n - F_n(\theta)$ are not only independent but identically distributed as some random variable W possessing a probability density function; that is,

$$P\{W_n \leq w\} = P\{W \leq w\} = \int_{-\infty}^{w} f(x)\,dx \qquad (n = 1, 2, \cdots).$$

The minimal assumptions on f, of course, are

$$\mathscr{E}W = 0, \qquad \mathscr{E}W^2 = \sigma^2 < \infty.$$

(Certain higher-order moments are presumed finite when we consider our methods of estimation.) We further suppose that

$$h(w) = -\frac{d \log f(w)}{dw}$$

exists (on all sets with positive probability) and that

$$\mathscr{E}h(W) = 0, \qquad \mathscr{E}h^2(W) = \lambda^2 > 0. \tag{5.5}$$

A sufficient condition for the former is that W be symmetrically distributed about the origin. The latter, it will be noted, excludes constant densities. We are also going to assume that f is independent of θ although, of course, this need not be the case for the validity of any of our results.

The likelihood of a realization $Y_1 = y_1, \cdots, Y_n = y_n$, when θ is the true parameter value, is simply

$$L_n(y_1, \cdots, y_n; \theta) = \prod_{k=1}^{n} f(y_k - F_k(\theta)), \qquad \theta \in \Omega.$$

Here Ω is an interval, finite or infinite, denoting the "natural" domain of the parameter, whereas J was a (not necessarily proper) subinterval of Ω defined by *a priori* knowledge. We have

$$\frac{\partial \log L_n}{\partial \theta} = \sum_{k=1}^{n} \dot{F}_k(\theta) h(y_k - F_k(\theta)), \tag{5.6}$$

which is a linear combination of independent, identically distributed random variables with coefficients depending on the unknown parameter. In view of our restriction, Equation 5.5, it follows that

$$I_n^2(\theta) = \mathscr{E}_\theta \left(\frac{\partial \log L_n}{\partial \theta}\right)^2 = \lambda^2 \sum_{k=1}^{n} \dot{F}_k^2(\theta), \tag{5.7}$$

where $\lambda = \lambda_f$ is independent of θ.

Now let $t_n = t_n(Y_1, \cdots, Y_n)$ denote a θ-estimate based on the first n observations (rather than $n - 1$ as previously). Under regularity conditions, the celebrated Cramér–Rao inequality states that

$$\mathscr{E}(t_n - \theta)^2 \geq b_n^2(\theta) + \frac{\{1 + [db_n(\theta)/d\theta]\}^2}{I_n^2(\theta)}, \tag{5.8}$$

where $b_n(\theta)$ is here the estimate's bias. The usual form in which the regularity conditions are written is (see, for example, Hodges and Lehmann, 1951) as follows:

i. Ω is open.
ii. $\partial \log L_n/\partial\theta$ exists for all $\theta \in \Omega$ and almost all points $y = (y_1, \cdots, y_n)$.
iii. $\mathscr{E}_\theta(\partial \log L_n/\partial\theta)^2 > 0$ for all $\theta \in \Omega$.
iv. $\int L_n\, dy = 1$ and $\int (t_n - \theta) L_n\, dy = b_n(\theta)$ may be differentiated under the (multiple) integral signs.

Our Equation 5.5 ensures Conditions ii and iii, and Condition iv holds because f does not depend on θ. We note that Conditions ii and iv imply

$$\mathscr{E}_\theta\, \partial \log L_n/\partial\theta = 0.$$

The ratio of the right-hand side of Equation 5.8 to the left-hand side is called the (fixed sample size) efficiency of t_n when θ is the true parameter point in Ω. As is known, a necessary and sufficient condition for an estimate t_n to be such that this ratio is unity for all $\theta \in \Omega$ is that t_n

be a sufficient estimate (a statement concerning $F_1, \cdots, F_n$ and f) and that $\partial \log g_n/\partial\theta = K_n(\theta)(t - \theta)$, where g_n is the density of t_n. The right-hand side of Equation 5.8 is only a lower bound on the mean-square estimation error; there exist problems where the uniform minimum variance of regular unbiased estimates exceeds $1/\mathrm{I}_n{}^2(\theta)$ at every θ.

Let us restrict our attention to Consistent Asymptotically Normal (abbreviated CAN) estimates of the value of θ specifying $\{F_n(\theta)\}$, that is, those for which

$$\frac{t_n - \theta}{\sigma_n(\theta)} \sim N(0, 1)$$

as $n \to \infty$, where $\sigma_n(\theta)$ is some positive sequence approaching zero with increasing sample size for any $\theta \in \Omega$. We assume that

$$\limsup_n I_n{}^2(\theta)\sigma_n{}^2(\theta) = \frac{1}{\eta_\theta} \tag{5.9}$$

exists (possibly as $+\infty$). Here η_θ is called the *asymptotic efficiency of* $\{t_n\}$ when θ is the true parameter value. If $\operatorname{Var} t_n \cong \sigma_n{}^2(\theta)$ and $db_n(\theta)/d\theta \to 0$ as $n \to \infty$ for all $\theta \in \Omega$, then it follows from Equation 5.8 that $\eta_\theta \leq 1$ for all $\theta \in \Omega$.

If a CAN estimate is such that $\eta_\theta = 1$ for all $\theta \in \Omega$, it is called a Consistent Asymptotically Normal Efficient (abbreviated CANE) estimate. This definition is made without restrictions entailing $\eta_\theta \leq 1$. CANE estimates sometimes do fail to have minimum asymptotic variance within the class of CAN estimates because the class is too broad to permit such a minimum. LeCam (1953), for example, has shown how to construct a set of superefficient estimates, that is, $\eta_\theta > 1$ for some $\theta \in \Omega$, from a given CANE estimate whose asymptotic variance obeys certain mild conditions. The basic idea is to define the new estimate in such a way that its bias goes to zero as $n \to \infty$ for all θ in Ω, but its derivative approaches a strictly negative number at isolated points in Ω. In other words, the lower bound in Equation 5.8 is attained and forced to be asymptotically smaller than $1/I_n{}^2(\theta)$ at some parameter values. The saving feature is that a parameter set of superefficiency must have Lebesque measure zero.

With these remarks as introductory material, we now compute the asymptotic efficiencies of the estimates which were the subject matter of Theorems 4.2′ and 5.1. In accordance with our initial discussion in this chapter concerning restrictions on $g_1, g_2, \cdots$, we impose the hypotheses of the latter theorem. We take this opportunity to write out in full these hypotheses (getting rid of some implications among them) and the results concerning the two types of estimation procedures in the case of a bounded prior knowledge interval J.

THEOREM 5.2

Let Y_n $(n = 1, 2, \cdots)$ be an observable sequence of independent random variables with common variance σ^2. Let $\mathscr{E} Y_n = F_n(\theta)$ be prescribed up to a parameter value θ which is known to be an interior point of a finite interval $J = [\xi_1, \xi_2]$. We impose the following conditions:

1. The derivative $\dot{F}_n$ exists and is one-signed on J for each n.
2. $B_n^2 = \sum_{k=1}^n b_k^2 \to \infty$ as $n \to \infty$, where $b_k = \inf_{x \in J} |\dot{F}_k(x)|$.
3. $\Sigma b_n^4/B_n^4 < \infty$.
4. Each $g_n(x) = |\dot{F}_n(x)|/b_n$ is continuous on J, and the sequence converges uniformly to a limit function $g(x)$. We set
$$c = \sup_{x \in J} g(x), \; 1 \le c < \infty.$$
5. $\sup_n \mathscr{E}[Y_n - F_n(\theta)]^{2p} < \infty$, where $p \ge 2$ is the smallest integer (assumed to exist) for which $\lim_n B_n^{p-2} b_n = +\infty$.

Let $t_1^{(i)} = \theta_0^{(i)}$ be fixed arbitrarily in J, and let

$$t_{n+1}^{(i)} = [t_n^{(i)} + a_n^{(i)}[Y_n - F_n(t_n^{(i)})]]_{\xi_1}^{\xi_2} \qquad (n = 1, 2, \cdots),$$

where

$$a_n^{(1)} = A_1 \operatorname{sgn} \dot{F}_n \frac{b_n}{B_n^2},$$
$$a_n^{(2)} = A_2 \frac{\dot{F}_n(t_n^{(2)})}{\sum_{k=1}^n \dot{F}_k^2(t_k^{(2)})},$$
$$a_n^{(3)} = A_3 \frac{\dot{F}_n(\theta_0^{(3)})}{\sum_{k=1}^n \dot{F}_k^2(\theta_0^{(3)})},$$

and the A's are positive constants. Then, as $n \to \infty$,

$$\sqrt{\sum_{k=1}^n \dot{F}_k^2(\theta)}\,(t_n^{(i)} - \theta)/\sigma \sim N(0, Q_i^2),$$

where

$$Q_1^2 = V(A_1 g(\theta)) \qquad \text{provided that} \quad A_1 > \tfrac{1}{2},$$
$$Q_2^2 = V(A_2) \qquad \text{provided that} \quad A_2 > c^2/2,$$
$$Q_3^2 = V\left(A_3 \frac{g(\theta)}{g(\theta_0^{(3)})}\right) \qquad \text{provided that} \quad A_3 > c^2/2,$$

and $V(x) = x^2/(2x - 1)$. The same limiting distribution obtains if $\dot{F}_k(\theta)$ is replaced by $\dot{F}_k(t_k^{(i)})$ in the norming sequence.

For $x \in J$, define

$$\Phi(x) = \int_{\xi_1}^{x} g(\xi)\, d\xi,$$

and let $\Psi = \Phi^{-1}$ be the inverse function. For y in $J^* = [0, \Phi(\xi_2)]$, define

$$F_n^*(y) = F_n(\Psi(y)).$$

Let $t_1^* = \varphi_0$ **be arbitrary in** J^***, and let**

$$t_{n+1}^* = \left[t_n^* + \frac{\dot{F}_n^*(\varphi_0)}{\sum_{k=1}^{n} \dot{F}_k^{*2}(\varphi_0)}[Y_n - F_n^*(t_n^*)]\right]_0^{\Phi(\xi_2)}$$

$$t_n = \Psi(t_n^*) \qquad (n = 1, 2, \cdots).$$

Then, as $n \to \infty$**,**

$$\sqrt{\sum_{k=1}^{n} \dot{F}_k^2(\theta)}\,(t_n - \theta)/\sigma \sim N(0, 1),$$

and the same holds true if $\dot{F}_k(\theta)$ **is replaced by** $\dot{F}_k(t_k)$ in the norming sequence.

We further now assume that the $Y_n - F_n(\theta)$ have a common probability destiny function f, which does not depend on θ, and derivative f', such that

$$\int_{-\infty}^{\infty} f'(w)\,dw = 0 \text{ and } \int_{-\infty}^{\infty} \frac{f'^2(w)}{f(w)}\,dw = \lambda^2 > 0.$$

Then the asymptotic efficiency of $\{t_n^{(i)}\}$ is

$$\eta_\theta^{(i)} = \frac{1}{\sigma^2\lambda^2 Q_i^2} \qquad (i = 1, 2, 3),$$

and the asymptotic efficiency of $\{t_n\}$ is

$$\eta = \frac{1}{\sigma^2\lambda^2} \leq 1$$

with equality when and only when f is the $N(0, \sigma^2)$ density.

Proof. From the uniform convergence of g_n to g on J, it follows that

$$\frac{d_n}{b_n} = \sup_J g_n(x) \to \sup_J g(x)$$

as $n \to \infty$. In other words, the numbers c defined in Theorem 4.2 and above in Condition 4 are indeed equal.

The formulas for the various asymptotic efficiencies immediately result from Equations 5.5, 5.7, 5.9, and the preceding expressions for the corresponding $\sigma_n^2(\theta)$'s.

The only statement requiring verification, then, is the final one. We may obviously assume λ^2 is finite. Then, by the Schwarz Inequality,

$$\sigma^2\lambda^2 = \int_{-\infty}^{\infty} w^2 f(w)\, dw \cdot \int_{-\infty}^{\infty} \frac{f'^2(w)}{f(w)}\, dw \geq \left|\int_{-\infty}^{\infty} wf'(w)\, dw\right|^2. \quad (5.10)$$

Since $w^2 f(w)$ is integrable, we must have $\lim_{|w|\to\infty} wf(w) = 0$, and therefore (assuming f is absolutely continuous),

$$\int_{-\infty}^{\infty} wf'(w)\, dw = wf(w)\Big|_{-\infty}^{+\infty} - \int_{-\infty}^{\infty} f(w)\, dw = -1.$$

This proves that $\sigma^2\lambda^2 \geq 1$. The necessary and sufficient condition for equality in Equation 5.10, that is, $\sigma^2\lambda^2 = 1$, is that the integrands $w^2 f(w)$ and $f'^2(w)/f(w)$ be linearly related:

$$f'(w) = Kwf(w) \qquad (-\infty < w < \infty).$$

This differential equation has a unique probability-density solution with first and second moments respectively equal to 0 and σ^2. It is

$$f(w) = \frac{1}{\sqrt{2\pi}\,\sigma} e^{-w^2/2\sigma^2},$$

as claimed. Q.E.D.

The transformed estimates $\{t_n\}$ in Theorem 5.2 are thus asymptotically efficient if and only if the errors of observation are Gaussian (and the untransformed if, in addition, the problem is asymptotically linear). Naturally the question arises as to how close $\eta = 1/\sigma^2\lambda^2 = \eta_f$ is to unity when f is "approximately" Gaussian.

Let us make this quantitative with Student's t-distribution having ν degrees of freedom:

$$f(w) = \frac{1}{\sqrt{\nu\pi}} \frac{\Gamma\left(\frac{\nu+1}{2}\right)}{\Gamma\left(\frac{\nu}{2}\right)} \left(1 + \frac{w^2}{\nu}\right)^{-(\nu+1)/2} \qquad (-\infty < w < \infty). \quad (5.11)$$

To meet Condition 5, we require that $\nu > 2p$, where $p \geq 2$ is an integer depending on the (unrelated) regression sequence. By symmetry, all odd-ordered moments are zero; in particular, $\mathscr{E}W = \mathscr{E}h(W) = 0$. The variance is $\sigma^2 = \nu/(\nu - 2)$ and, as $\nu \to \infty$, Equation 5.11 approaches the $N(0, 1)$ density function. After a somewhat lengthy but straightforward calculation, we obtain the formula

$$\eta = \frac{(\nu - 2)^2(\nu + 3)}{2(\nu + 1)(\nu + 2)}.$$

For large ν, $\eta = 1 - (4/\nu) + O(1/\nu^2)$. As the following table shows, the approach is not too rapid.

Degrees of Freedom in Student's Distribution	Asymptotic Efficiency of Transformed Estimates
5	0.34
6	0.43
7	0.50
8	0.55
10	0.63
15	0.75
20	0.81
25	0.84
50	0.89
100	0.94
400	0.99

For the Laplacian density

$$f(w) = \frac{1}{\sqrt{2}\,\sigma} e^{-\sqrt{2}|w|/\sigma} \qquad (-\infty < w < \infty) \tag{5.12}$$

all moments exist, so Condition 5 holds in every case. The odd-ordered moments are 0, and σ^2 is the variance. We find that

$$\lambda^2 = \int_{-\infty}^{\infty} \left[\frac{d \log f(w)}{dw}\right]^2 f(w)\, dw = \frac{2}{\sigma^2},$$

and hence

$$\eta = \tfrac{1}{2}.$$

It is to be noted for this density that

$$\mathscr{E}h'(W) = -\int_{-\infty}^{\infty} \frac{d^2 \log f(w)}{dw^2} f(w)\, dw = 0.$$

Consequently, by Equations 5.6 and 5.7,

$$-\mathscr{E}_\theta \frac{\partial^2 \log L_n}{\partial \theta^2} \neq \mathscr{E}_\theta \left(\frac{\partial \log L_n}{\partial \theta}\right)^2 \tag{5.13}$$

because the left-hand side is identically zero. But the assumption of equality in Equation 5.13 for all $\theta \in \Omega$ is fundamental to the proof that there is a solution of $\partial \log L_n/\partial\theta = 0$, that is, a ML estimate, which is a CANE estimate. (The function $h(\cdot)$ is a step function in this case.) Hence, the low asymptotic efficiency of our transformed estimate is only apparent.

5.4 Increased Efficiency

If the (non-Gaussian) noise density f is prescribed and such that

$$\mathscr{E}h'(W) = \mathscr{E}h^2(W),$$

the asymptotic efficiency of any of our methods can be improved whenever there is a CANE solution, $\hat{\theta}_n$, which makes Equation 5.6 vanish. (This generally imposes further conditions on the regression functions as well as on f. See, for example, LeCam, 1953.) If we now let t_n denote any of the recursive estimates listed in Theorem 5.2, the quantities

$$\hat{\hat{\theta}}_n = t_n + \frac{1}{\lambda^2} \frac{\sum_{k=1}^{n} \dot{F}_k(t_n) h(Y_k - F_k(t_k))}{\sum_{k=1}^{n} \dot{F}_k^2(t_k)}$$

will then have the same optimum large-sample statistical properties as the ML estimates. The fact that these require infinite memory, via the quantities $\dot{F}_k(t_n)$ $(k = 1, 2, \cdots, n)$, violates our ground rule that we restrict consideration to computationally feasible estimation schemes.

5.5 Large-Sample Confidence Intervals

Under the conditions of Theorem 5.2, we have shown, for the transformed estimates $\{t_n\}$, that

$$\lim_n P\{S_n(t_n - \theta) < \sigma x\} = \frac{1}{\sqrt{2\pi}} \int_{-\infty}^{x} \exp(-\tfrac{1}{2}\xi^2)\, d\xi$$

for every real number x. The norming sequence

$$S_n^2 = \sum_{k=1}^{n} \dot{F}_k^2(t_k) \tag{5.14}$$

is free of unknowns. The value of $\sigma^2 = \sigma_\theta^2$ can be consistently estimated, without further assumptions, by

$$s_n^2 = \frac{1}{n} \sum_{k=1}^{n} [Y_k - F_k(t_k)]^2,$$

and we can therefore set up large-sample confidence intervals for θ. Specifically, given a (small) number α, let c_α be such that $100(1 - \alpha)$ percent of the area under the standardized normal curve lies between $-c_\alpha$ and c_α. Then

$$t_n \pm c_\alpha \frac{s_n}{S_n} \tag{5.15}$$

is an asymptotic confidence interval on θ, having confidence coefficient $1 - \alpha$. That is to say, the probability that this interval will cover the unknown θ tends to $1 - \alpha$ as $n \to \infty$. For any (large) fixed n, there is, of course, positive probability that the interval in Equation 5.15 will not be a subset of J. In practice, the confidence interval will be taken as the intersection of the two.

5.6 Choice of Indexing Sequence

In some problems the integer-valued indexing of the regression functions results from prior selection of a set of discrete values of a continuous index (usually denoting "where" or "when" we sample). Suppose, then, we start with a prescribed function $F(\tau, \theta)$ of two continuous variables. For a particular choice $\tau_1, \tau_2, \cdots$ we obtain a sequence of mean values

$$F_k(\theta) = F(\tau_k, \theta) \qquad (\theta \in J; k = 1, 2, \cdots)$$

known up to the parameter θ. Generally, there will be no unique sequence $\tau_1, \tau_2, \cdots$ for which the resulting $\{F_k(\cdot)\}$ satisfies Conditions 1 through 5 of Theorem 5.2. The most obvious criterion for choosing such a sequence is maximization of each of the summands in Equation 5.14. That is, we define $\tau_k = \tau(t_k)$ by

$$\max_{\tau} \dot{F}^2(\tau, t_k) = \dot{F}^2(\tau_k, t_k), \tag{5.16}$$

where t_k is determined, in accordance with one of our iterative schemes, by $t_1, \cdots, t_{k-1}$ and $\tau_1, \cdots, \tau_{k-1}$. The range of τ values (it may be vector-valued) over which the maximum is taken will usually be restricted by further considerations.

A (trivial) example would be

$$F(\tau, \theta) = (\tau - \theta)^3 \text{ and } J = [0, 1].$$

The squared derivative with respect to θ is maximized over $\tau \in J$ by

$$\tau(\theta) = \begin{cases} 1 & \text{if } 0 < \theta \le \frac{1}{2}, \\ 0 & \text{if } \frac{1}{2} \le \theta < 1. \end{cases}$$

There is, then, a single regression function

$$F(x) = \begin{cases} (1 - x)^3, & 0 \le x \le \frac{1}{2}, \\ x^3, & \frac{1}{2} \le x \le 1, \end{cases}$$

with $b_n = \frac{3}{4}$ for all n.

5.7 A Single-Parameter Estimation Problem

Example 5.1. The following is a single-parameter specialization of a multiparameter problem. We seek to estimate recursively, from range measurements corrupted by additive noise, the true initial range to a body traveling in an inverse square law central force field with total system energy $E = 0$, that is, along a parabolic path. Solution of the more realistic multiparameter estimation problem for an elliptic path ($E < 0$) is worked out as Example 8.5 in Chapter 8.

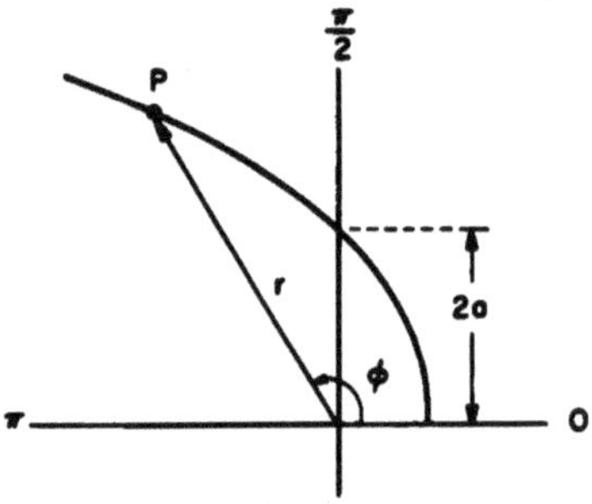

Figure 5.1 Trajectory of object

The polar coordinates (r, φ) of the parabola with focus at the origin shown in Figure 5.1 are related by

$$r = \frac{2a}{1 + \cos\varphi}, \tag{5.17}$$

where the angle φ is measured as indicated. If a force $f = -k/r^2$ is exerted by the origin on the point P, with (reduced) mass m, then

$$a = \frac{l^2}{2mk},$$

wherein

$$l = mr^2 \frac{d\varphi}{dt} = \text{const} \tag{5.18}$$

is the conserved angular momentum. The motion is thus determined by the values of three parameters: m, k, and l (plus an initial time).

We select

$$\theta = a$$

as the one to be estimated and presume the others given. We assume that at time $t = 0$ the coordinates of P are $(a, 0)$, that is, that the turning angle, which orients the axis of the parabola to the observational coordinate system, is also known. Integration of Equation 5.18 with r given by Equation 5.17 then yields the cubic equation

$$z + \tfrac{1}{3}z^3 = K\frac{t}{\theta^2} \qquad \left(K = \frac{l}{2m}\right) \tag{5.19}$$

for

$$z = \tan \tfrac{1}{2}\varphi. \tag{5.20}$$

There is a single positive root, namely,

$$z = z(t/\theta^2) = (u + \sqrt{1 + u^2})^{1/3} + (u - \sqrt{1 + u^2})^{1/3}$$
$$u = \frac{3K}{2} \cdot \frac{t}{\theta^2} \qquad (t \geq 0). \tag{5.21}$$

By Equations 5.17 and 5.20 the regression at time t is thus

$$r = \theta(1 + z^2) = F(t, \theta), \tag{5.22}$$

wherein z depends nonlinearly on t/θ^2 in accordance with Equation 5.21.

In the following we introduce a sequence of regressions $F_k(\theta) = F(\tau_k, \theta)$ by selecting appropriate observation times $0 < \tau_1 < \tau_2 < \cdots$, but for the time being we can continue to work with the continuous time variable. Furthermore, rather than introduce more symbols, we use θ as the dummy variable, where

$$0 < \xi_1 < \theta < \xi_2 < \infty,$$

and the end points are given.

Letting a dot denote differentiation with respect to the parameter (and not, as is customary, with respect to time), we have from Equation 5.22

$$\dot{F}(t, \theta) = 1 + z^2 + 2\theta z \frac{\partial z}{\partial \theta}.$$

But from Equation 5.19 we have

$$\frac{\partial z}{\partial \theta}(1 + z^2) = -2K\frac{t}{\theta^3},$$

so that

$$\dot{F}(t, \theta) = 1 + z^2 - 4K\frac{t}{\theta^2} \cdot \frac{z}{1 + z^2}.$$

Returning to z, by using Equation 5.19 once again, we find that

$$\dot{F}(t, \theta) = -H(z^2), \qquad H(x) = \frac{x^2 + 6x - 3}{3(1 + x)}. \tag{5.23}$$

This expression, together with Equation 5.21, is the basis for all further considerations.

The quadratic numerator in H vanishes at $x = -3 - 2\sqrt{3} < 0$ and $-3 + 2\sqrt{3} > 0$. Since $H(0) = -1$, it follows that

$$H(x) > 0 \qquad \text{for all } x > x_0 = -3 + 2\sqrt{3} = 0.464. \tag{5.24}$$

In addition, $H(x)$ increases monotonically with $x > x_0$.

Now, the solution $z = \tan \frac{1}{2}\varphi$ increases with t for every fixed θ, that is, every fixed path. But $z = z(t/\theta^2)$, so z must decrease with increasing θ for every fixed t. In particular,

$$z(t/\theta^2) > z(t/\xi_2^{\,2}).$$

If we now define t_0 as the positive time at which

$$z^2(t_0/\xi_2^{\,2}) = x_0, \tag{5.25}$$

we will then have

$$z^2(t/\theta^2) > x_0.$$

Consequently, by Equations 5.23 and 5.24, we obtain

$$\dot{F}(t, \theta) < 0 \qquad \text{for all } t > t_0 \text{ and all } \theta \in J.$$

Thus, Condition 1 of Theorem 5.2 will be met if we begin our observations at any time after t_0, which is defined by Equation 5.25 in terms of ξ_2 and K.

We next examine the behavior of $|\dot{F}(t, \theta)|$ as $t \to \infty$ for θ in J. From Equation 5.21, we see that

$$\lim_{t \to \infty} \frac{z(t/\theta^2)}{t^{1/3}} = \left(\frac{3K}{\theta^2}\right)^{1/3},$$

and, from Equation 5.23,

$$\lim_{x \to \infty} \frac{H(x)}{x} = \frac{1}{3}.$$

Consequently,

$$|\dot{F}(t, \theta)| \simeq \frac{1}{3}\left(\frac{3K}{\theta^2}\right)^{2/3} t^{2/3} = A(\theta)t^{2/3} \tag{5.26}$$

as $t \to \infty$. Furthermore, with $t > t_0$, we obtain

$$b(t) = \inf_{\theta \in J} |\dot{F}(t, \theta)| = \inf_{\theta \in J} H\left(z^2\left(\frac{t}{\theta^2}\right)\right) = H\left(z^2\left(\frac{t}{\xi_2^{\,2}}\right)\right),$$

and it is easy to see that

$$g(t, \theta) = \frac{|\dot{F}(t, \theta)|}{b(t)} \to \frac{A(\theta)}{A(\xi_2)} = g(\theta)$$

uniformly on J as $t \to \infty$. Thus, the functions $g_k(\theta) = g(\tau_k, \theta)$ $(k = 1, 2, \cdots)$ will satisfy Condition 4 for any sequence

$$t_0 < \tau_1 < \tau_2 < \cdots$$

increasing to infinity.

There are many such sequences for which Conditions 2 and 3 are met. For instance, we can take slowly increasing times such as

$$\tau_k = \log^{3\alpha/4} k \qquad (\text{any } \alpha > 0, k = 1, 2, \cdots).$$

Then, by Equation 5.26, we will have for $n \to \infty$

$$\sum_{k=1}^{n} \dot{F}^2(\tau_k, \theta) \cong A^2(\theta) \sum_{k=1}^{n} \log^\alpha k. \tag{5.27}$$

According to Sacks' (1958) Lemma 4,

$$\sum_{k=1}^{n} \log^\alpha k \cong n \log^\alpha n. \tag{5.28}$$

Thus, as $n \to \infty$,

$$b_n^2 \sim \log^\alpha n, \qquad B_n^2 \sim n \log^\alpha n,$$

and both Conditions 2 and 3 hold. In addition, Condition 5 is true if the additive noise in the range measurements has finite fourth-order moments.

PART II

THE VECTOR-PARAMETER CASE

6. Mean-Square and Probability-One Convergence

In this chapter we turn our attention to the more realistic situation where the regression function is known except for a number of scalar parameters; or equivalently, if there are $p > 1$ unknowns, a p-dimensional vector parameter. We will study the quadratic-mean and almost sure convergence, in that order, of (column) vector estimator sequences of the form

$$\mathbf{t}_{n+1} = \mathbf{t}_n + \mathbf{a}_n[Y_n - F_n(\mathbf{t}_n)] \qquad (n = 1, 2, \cdots; t_1 \text{ arbitrary}). \quad (6.1)$$

The scalar observable is

$$Y_n = F_n(\boldsymbol{\theta}) + W_n,$$

where $\{F_n(\cdot)\}$ is known, $\boldsymbol{\theta}$ is the p-dimensional vector to be estimated, W_n is the residual error, and $\{\mathbf{a}_n\}$ is a suitably chosen sequence of p-dimensional gain vectors. Owing to its considerable complexity, the question of a large-sample distribution for the vector estimates is not examined in this monograph. However, the technique of analysis used in Chapters 3 and 4 would be a logical starting point if we were to consider this problem.

Our approach to the vector-parameter case is patterned after the one-dimensional treatment found in Chapter 2. We linearize the regression, assuming the existence of first partials, by invoking the mean-value theorem:

$$F_n(\mathbf{t}_n) = F_n(\boldsymbol{\theta}) + \dot{\mathbf{F}}_n'(\mathbf{u}_n)(\mathbf{t}_n - \boldsymbol{\theta}),$$

where

$\mathbf{u}_n$ lies on the line segment joining $\mathbf{t}_n$ and $\boldsymbol{\theta}$,
$\dot{\mathbf{F}}_n(\mathbf{u}_n)$ is the gradient of F_n evaluated at $\mathbf{u}_n$, and
$\dot{\mathbf{F}}_n'(\mathbf{u}_n)$ is the (row vector) transpose of the (column vector) $\dot{\mathbf{F}}_n(\mathbf{u}_n)$.

From Equation 6.1 we then have

$$\mathbf{t}_{n+1} - \boldsymbol{\theta} = [\mathbf{I} - \mathbf{a}_n\dot{\mathbf{F}}_n'(\mathbf{u}_n)](\mathbf{t}_n - \boldsymbol{\theta}) + \mathbf{a}_n W_n, \tag{6.2}$$

where $\mathbf{I}$ is the $p \times p$ identity matrix. Iterating this back to $n = 1$ gives

$$\mathbf{t}_{n+1} - \boldsymbol{\theta} = \prod_{j=1}^{n} [\mathbf{I} - \mathbf{a}_j\dot{\mathbf{F}}_j'(\mathbf{u}_j)](\mathbf{t}_1 - \boldsymbol{\theta}) + \sum_{j=1}^{n} \prod_{i=j+1}^{n} [\mathbf{I} - \mathbf{a}_i\dot{\mathbf{F}}_i'(\mathbf{u}_i)]\mathbf{a}_j W_j, \tag{6.3}$$

where, both now and later, $\prod_{j=m}^{n} \mathbf{A}_j$ means the matrix product $\mathbf{A}_n\mathbf{A}_{n-1} \cdots \mathbf{A}_m$ (i.e., the product is to be read "backward").

It is clear from Equation 6.3 that the large-sample statistical properties of $\{\mathbf{t}_n - \boldsymbol{\theta}\}$ are crucially dependent on the asymptotic (large n) properties of iterated transformations of the type

$$\mathbf{P}_n = \prod_{j=1}^{n} (\mathbf{I} - \mathbf{a}_j\mathbf{h}_j'), \tag{6.4}$$

where $\mathbf{a}_j$ and $\mathbf{h}_j'$ are p-dimensional column and row vectors, respectively. We begin by studying conditions on deterministic sequences of p-vectors, $\{\mathbf{a}_j\}$ and $\{\mathbf{h}_j\}$, which are sufficient to guarantee that $\mathbf{P}_n$ converges to zero (that is, the null matrix) as $n \to \infty$.

In the one-dimensional case, this problem is trivial: P_n converges to zero if the positive a_jh_j's are such that $\sum_j a_jh_j = \infty$ and $a_jh_j > 1$ only finitely often. (This was so because of the scalar inequality $1 - x \le e^{-x}$.) In higher dimensions, life is not so simple, and we must think in terms of matrix eigenvalues. In what follows, we make use of the following statement.

Definition of norm. For a square symmetric matrix $\mathbf{P}$, let $\lambda_{\min}(\mathbf{P})$ and $\lambda_{\max}(\mathbf{P})$, respectively, denote the smallest and largest eigenvalues of $\mathbf{P}$ (all of which are real). For any rectangular matrix $\mathbf{P}$, we use as its norm

$$\|\mathbf{P}\| = [\lambda_{\max}(\mathbf{P}'\mathbf{P})]^{1/2},$$

where $\mathbf{P}'$ is the transpose of $\mathbf{P}$.

As so defined, $\|\cdot\|$ is bona fide norm since we are concerned only with real matrices. If $\mathbf{P}$ is a $p \times 1$ matrix (i.e., a column), then $\|\mathbf{P}\| = (\mathbf{P}'\mathbf{P})^{1/2}$, the familiar Euclidean length. If $\mathbf{P}$ is a $1 \times p$ matrix (i.e., a row), then

$\mathbf{P'P}$ is a $p \times p$ matrix. It has but one nonzero eigenvalue; namely, $\mathbf{PP'} = \|\mathbf{P}\|^2$, with, incidentally, an associated right-sided eigenvector $\mathbf{P'}$. Finally, if $\mathbf{P}$ is of the form $\mathbf{ah'}$, where $\mathbf{a}$ and $\mathbf{h}$ are column vectors, then $\mathbf{P'P} = \mathbf{h(a'a)h'} = \|\mathbf{a}\|^2\mathbf{hh'}$; therefore, we have

$$\|\mathbf{P}\|^2 = \|\mathbf{a}\|^2\lambda_{\max}(\mathbf{hh'}) = \|\mathbf{a}\|^2\|\mathbf{h}\|^2.$$

As is generally the case with matrix norms,

$$\|\mathbf{PQ}\| \le \|\mathbf{P}\|\,\|\mathbf{Q}\|,$$
$$\|\mathbf{P} + \mathbf{Q}\| \le \|\mathbf{P}\| + \|\mathbf{Q}\|,$$

provided that $\mathbf{P}$ and $\mathbf{Q}$ are such that the operations are defined.

It is evident that a sequence of matrices $\{\mathbf{P}_n\}$ converges to the matrix of zeros as $n \to \infty$ if and only if $\|\mathbf{P}_n\| \to 0$. In the particular case where $\mathbf{P}_n$ is given by Equation 6.4, we are tempted to make use of the (correct) inequality

$$\|\mathbf{P}_n\| \le \prod_{j=1}^{n} \|\mathbf{I} - \mathbf{a}_j\mathbf{h}_j'\|$$

in order to find conditions on the vector sequences $\{\mathbf{a}_j\}$ and $\{\mathbf{h}_j\}$ which will ensure $\|\mathbf{P}_n\| \to 0$. This approach proves to be fruitless. In fact, it can be shown that

$$\|\mathbf{I} - \mathbf{ah'}\| \ge 1$$

for any "elementary matrix" of the form $\mathbf{I} - \mathbf{ah'}$, where $\mathbf{a}$ and $\mathbf{h}$ are column vectors, so the above-cited inequality tells us nothing about the convergence of $\|\mathbf{P}_n\|$.

The successful approach involves grouping successive products together and exploring an inequality of the form

$$\|\mathbf{P}_n\| \le \prod_{k=1}^{k(n)} \Big\|\prod_{j \in J_k} (\mathbf{I} - \mathbf{a}_j\mathbf{h}_j')\Big\|,$$

where J_k is a set of consecutive integers. This idea is the basis of the following theorem.

6.1 Theorem Concerning Divergence to Zero of Products of Elementary Matrices and Assumptions (B1 Through B5)

THEOREM 6.1

Let $\{\mathbf{a}_j\}$ and $\{\mathbf{h}_j\}$ be a sequence of p-vectors and define

$$\mathbf{P}_n = \prod_{j=1}^{n} (\mathbf{I} - \mathbf{a}_j\mathbf{h}_j')$$

for all $n \geq 1$. Then we have

$$\lim_n \|\mathbf{P}_n\| = 0$$

if the following assumptions hold:

B1. $\|\mathbf{a}_n\| \, \|\mathbf{h}_n\| \to 0$ as $n \to \infty$.

B2. $\sum_n \|\mathbf{a}_n\| \, \|\mathbf{h}_n\| = \infty$.

B3. There exists a sequence of integers $1 = \nu_1 < \nu_2 < \nu_3 \cdots$ such that, with $p_k = \nu_{k+1} - \nu_k$, we have

$$p \leq p_k \leq q < \infty \qquad (k = 1, 2, \cdots)$$

and

$$\liminf_k \frac{1}{p_k} \lambda_{\min} \left(\sum_{j \in J_k} \frac{\mathbf{h}_j \mathbf{h}_j'}{\|\mathbf{h}_j\|^2} \right) = \tau^2 > 0,$$

where both now and later J_k is the index set

$$\{\nu_k, \nu_k + 1, \cdots, \nu_{k+1} - 1\}.$$

B4. $$\limsup_k \frac{\max_{j \in J_k} \|\mathbf{a}_j\| \, \|\mathbf{h}_j\|}{\min_{j \in J_k} \|\mathbf{a}_j\| \, \|\mathbf{h}_j\|} = \rho < \infty.$$

B5. $$\liminf_n \frac{\mathbf{a}_n' \mathbf{h}_n}{\|\mathbf{a}_n\| \, \|\mathbf{h}_n\|} > \alpha = \sqrt{\frac{1 - \tau^2}{1 - \tau^2 + (\tau/\rho)^2}}.$$

6.2 Discussion of Assumptions and Proof

Before embarking on the proof of this fundamental result, let us try to give some insight into the meaning of the assumptions. The first, second, and fourth are assumptions concerning the rate of decay of the product $\|\mathbf{a}_n\| \|\mathbf{h}_n\|$. The first two are particularly in the spirit of their one-dimensional analogues.

Assumption B3 has the following interpretation. For any set of p-dimensional unit vectors $\mathbf{u}_1, \mathbf{u}_2, \cdots, \mathbf{u}_r$, we have

$$\lambda_{\min} \left(\sum_{j=1}^{r} \mathbf{u}_j \mathbf{u}_j' \right) = \min_{\|\mathbf{x}\|=1} \mathbf{x}' \left(\sum_{j=1}^{r} \mathbf{u}_j \mathbf{u}_j' \right) \mathbf{x} = \min_{\|\mathbf{x}\|=1} \sum_{j=1}^{r} (\mathbf{u}_j' \mathbf{x})^2.$$

Now $|(\mathbf{u}_j'\mathbf{x})|$ is precisely the distance from $\mathbf{u}_j$ to the hyperplane through the origin whose unit normal is $\mathbf{x}$; therefore, $\sum_{j=1}^{r} (\mathbf{u}_j'\mathbf{x})^2 = d^2(\mathbf{x})$ is the sum of the squared distances from the $\mathbf{u}_j$'s to that hyperplane. Since $d^2(\mathbf{x})$ is continuous in $\mathbf{x}$, it actually achieves its minimum on the (compact) surface of the unit sphere. Thus, the value of

$$\lambda_{\min}\left(\sum_{j=1}^{r} \mathbf{u}_j\mathbf{u}_j'\right) = \inf_{\|\mathbf{x}\|=1} d^2(\mathbf{x})$$

is the sum of the squared distances from the $\mathbf{u}_j$'s to the particular $(p - 1)$-dimensional hyperplane that "best fits" the vector set $\mathbf{u}_1, \cdots, \mathbf{u}_r$. Assumption B3, therefore, requires that the normalized vector sequence $\mathbf{h}_1/\|\mathbf{h}_1\|, \mathbf{h}_2/\|\mathbf{h}_2\|, \cdots$ can be broken up into groups of finite size, the kth group containing $r = p_k$ members, in such a way that the average squared distance from the vectors $\mathbf{h}_j/\|\mathbf{h}_j\|$ $(j \in J_k)$ to any $(p - 1)$-dimensional hyperplane remains bounded away from zero as $k \to \infty$.

Loosely speaking, the sequence $\mathbf{h}_1, \mathbf{h}_2, \cdots$ must therefore "repeatedly span" Euclidean p-space. No direction must be neglected. This makes good intuitive sense. Indeed, let $\mathbf{x}$ be a generic point in p-space and set $\mathbf{x}_{n+1} = \mathbf{P}_n\mathbf{x}$. Then we have

$$\mathbf{x}_{n+1} = \mathbf{x}_n - (\mathbf{h}_n'\mathbf{x}_n)\mathbf{a}_n;$$

that is, $\mathbf{x}_{n+1}$ is obtained from $\mathbf{x}_n$ by subtracting off a piece of $\mathbf{x}_n$ pointing in the direction $\mathbf{a}_n$. If $\mathbf{P}_n$ is to map every $\mathbf{x}$ into the origin as $n \to \infty$, as it must if $\|\mathbf{P}_n\| \to 0$, then $\mathbf{h}_n$ must have at least a nonzero projection on all p-coordinate axes infinitely often as $n \to \infty$. Assumption B3 requires just this, in a precise way.

There is also a relationship that exists between τ^2 and the limiting value of the ratio of the largest to smallest eigenvalue of the matrix $\sum_{k=1}^{n} \mathbf{h}_k\mathbf{h}_k'$ (which is the subject matter of Lemma 7b of the Appendix). This ratio, sometimes called the conditioning number, is a measure of how close a nonsingular matrix is to being singular.

In the scalar-parameter case, we required that the gain have the same, presumed constant, sign as the regression-function derivative. In the present notation, the requirement would read

$$\frac{a_n h_n}{|a_n|\,|h_n|} = \operatorname{sgn} a_n \cdot \operatorname{sgn} h_n > 0$$

for all sufficiently large n. In higher dimensions, the natural analogue of the product numerator is the inner product, and of the absolute values, the lengths. Therefore, we must ensure that

$$\liminf_{n} \mathbf{a}_n'\mathbf{h}_n/\|\mathbf{a}_n\|\,\|\mathbf{h}_n\| > 0,$$

and it might seem reasonable that this is sufficient. But Assumption B5 demands much more of the cosines of the angles between the $\mathbf{a}_n$'s and $\mathbf{h}_n$'s. It requires that their smallest limit point be strictly larger than a certain positive quantity α, which depends on every member of both the sequences $\{\mathbf{a}_n\}$ and $\{\mathbf{h}_n\}$. (We note $\tau^2 \leq 1$ is always the case, so $0 \leq \alpha < 1$, as should be.) Moreover, *the lower bound in Assumption B5 is an essential one*. This is graphically demonstrated by the following example in which Assumptions B1 through B4 hold,

$$\liminf_n \mathbf{a}_n'\mathbf{h}_n/\|\mathbf{a}_n\|\ \|\mathbf{h}_n\| = \alpha > 0$$

(so that Assumption B5 is "just barely" violated), but $\mathbf{P}_n$ does not converge to zero.

Example 6.1. We take

$$\mathbf{a}_n = \frac{1}{n}\begin{bmatrix}\cos\varphi \\ \sin\varphi\end{bmatrix}, \qquad \mathbf{h}_n = \begin{cases}\begin{bmatrix}1\\0\end{bmatrix} & \text{if } n \text{ is odd,} \\[2ex] \begin{bmatrix}0\\1\end{bmatrix} & \text{if } n \text{ is even,}\end{cases}$$

where $0 < \varphi < \pi/2$. Assumptions B1 and B2 are immediate because $\|\mathbf{a}_n\| = 1/n$ and $\|\mathbf{h}_n\| = 1$. The limit inferior in Assumption B5 is simply

$$\liminf_n \mathbf{a}_n'\mathbf{h}_n/\|\mathbf{a}_n\|\ \|\mathbf{h}_n\| = \min(\cos\varphi, \sin\varphi) \leq 1/\sqrt{2},$$

with equality only at $\varphi = \pi/4$. With regard to Assumption B4, we have

$$\rho = \limsup_k \frac{\max\limits_{\nu_k \leq j \leq \nu_{k+1}-1} \|\mathbf{a}_j\|\ \|\mathbf{h}_j\|}{\min\limits_{\nu_k \leq j \leq \nu_{k+1}-1} \|\mathbf{a}_j\|\ \|\mathbf{h}_j\|} = \lim_k \left(\frac{\nu_{k+1} - 1}{\nu_k}\right) = 1$$

for any strictly increasing integer sequence $\{\nu_k\}$ whose successive differences remain bounded. For the particular choice $\nu_k = 2k - 1$, we have

$$\sum_{j=2k-1}^{2k} \mathbf{h}_j\mathbf{h}_j' = \begin{bmatrix}1 & 0\\0 & 1\end{bmatrix};$$

and therefore

$$\frac{1}{p_k}\lambda_{\min}\left(\sum_{j\in J_k} \frac{\mathbf{h}_j\mathbf{h}_j'}{\|\mathbf{h}_j\|^2}\right) = \frac{1}{2}$$

identically in k. It can be further shown that the value of τ^2 in Assumption B3 cannot exceed $\frac{1}{2}$ for any choice of indices. In other words, $\alpha = \sqrt{1 - \tau^2} \geq 1/\sqrt{2}$ is true in all cases, and Assumption B5 will be even more violated if $p_k > 2$. We now take $\varphi = \pi/4$, which gives

$$\liminf_n \frac{\mathbf{a}_n'\mathbf{h}_n}{\|\mathbf{a}_n\|\ \|\mathbf{h}_n\|} = \frac{1}{\sqrt{2}} = \alpha.$$

With this choice of the angle, we thereby satisfy Assumptions B1 through B4 and violate Assumption B5 to the extent that equality rather than inequality holds. But, for any φ, we have

$$(\mathbf{I} - \mathbf{h}_j\mathbf{a}_j')\begin{bmatrix}\sin\varphi \\ -\cos\varphi\end{bmatrix} = \begin{bmatrix}\sin\varphi \\ -\cos\varphi\end{bmatrix} = \boldsymbol{\xi},$$

whether j is odd or even. Thus, we see that

$$\mathbf{P}_n'\boldsymbol{\xi} = \boldsymbol{\xi}$$

for all $n \geq 1$, and we have exhibited a (nontrivial) fixed point of every one of the transformations. Therefore, $\{\mathbf{P}_n\}$ cannot tend to the zero matrix.

In particular, then, the lower bound in Assumption B5 cannot generally be replaced by zero, in contrast to what we might have expected by analogy with the scalar-parameter case. We now prove the theorem.

Proof of Theorem 6.1. From Assumption B1 it follows that $\sup_n \|\mathbf{I} - \mathbf{a}_n\mathbf{h}_n'\| = M < \infty$. For any n, let $K = K(n)$ be the largest integer such that $\nu_K \leq n$, so that $\nu_K \leq n \leq \nu_{K+1} - 1$. Then we have

$$\mathbf{P}_n = \prod_{j=\nu_K}^{n} (\mathbf{I} - \mathbf{a}_j\mathbf{h}_j')\mathbf{P}_{\nu_K - 1}$$

and, by Assumption B3,

$$\|\mathbf{P}_n\| \leq \prod_{j=\nu_K}^{n} \|\mathbf{I} - \mathbf{a}_j\mathbf{h}_j'\| \, \|\mathbf{P}_{\nu_K-1}\| \leq M^q\|\mathbf{P}_{\nu_K-1}\|.$$

It therefore suffices to show that

$$\mathbf{P}_{\nu_K - 1} = \prod_{k=1}^{K-1} \prod_{j\in J_k} (\mathbf{I} - \mathbf{a}_j\mathbf{h}_j') \to \mathbf{0} \tag{6.5}$$

as K tends to infinity with n over some subset of the positive integers.

To do this, we set

$$\mathbf{Q}_k = \prod_{j\in J_k} (\mathbf{I} - \mathbf{a}_j\mathbf{h}_j'), \qquad \mathbf{T}_k = \frac{1}{\Delta_k}\sum_{j\in J_k} \mathbf{a}_j\mathbf{h}_j', \tag{6.6}$$

where

$$\Delta_k = \Big(\sum_{j\in J_k} \|\mathbf{a}_j\|^2\|\mathbf{h}_j\|^2\Big)^{1/2}. \tag{6.7}$$

By virtue of Assumptions B1 and B3, we have

$$\limsup_k \Delta_k^2 \leq q \limsup_k \|\mathbf{a}_k\|^2\|\mathbf{h}_k\|^2 = 0, \tag{6.8}$$

where, unless otherwise noted, k runs over all positive integers. It is not difficult to see that

$$\mathbf{Q}_k'\mathbf{Q}_k = \mathbf{I} - \Delta_k(\mathbf{T}_k + \mathbf{T}_k') + \Delta_k^2\mathbf{E}_k$$

for some matrix $\mathbf{E}_k$, whose norm is uniformly bounded in k. Thus, since the matrices are symmetric,

$$\begin{aligned}\|\mathbf{Q}_k\|^2 = \|\mathbf{Q}_k'\mathbf{Q}_k\| &\leq \|\mathbf{I} - \Delta_k(\mathbf{T}_k + \mathbf{T}_k')\| + O(\Delta_k^2)\\ &= \lambda_{\max}[\mathbf{I} - \Delta_k(\mathbf{T}_k + \mathbf{T}_k')] + O(\Delta_k^2)\\ &= 1 - \Delta_k\lambda_{\min}(\mathbf{T}_k + \mathbf{T}_k') + O(\Delta_k^2)\end{aligned} \tag{6.9}$$

as $k \to \infty$. Consequently, if we can show that

$$\liminf_k \lambda_{\min}(\mathbf{T}_k + \mathbf{T}_k') > 3c > 0, \tag{6.10}$$

for some such number c, we are done. For then, since $\Delta_k \to 0$, from Equations 6.9 and 6.10 we have

$$0 \leq \|\mathbf{Q}_k\|^2 \leq 1 - 2c\Delta_k$$

(say) for all large enough k. But $1 - x \leq e^{-x}$ is always true, so that

$$\prod_{k=N}^{n} \|\mathbf{Q}_k\| \leq \exp\left\{-c \sum_{k=N}^{n} \Delta_k\right\}. \tag{6.11}$$

Since the square root of the sum of squares is never smaller than the sum of the absolute values,

$$\sum_k \Delta_k \geq \sum_k \sum_{j \in J_k} \|\mathbf{a}_j\|\,\|\mathbf{h}_j\| = \sum_k \|\mathbf{a}_k\|\,\|\mathbf{h}_k\| = \infty \tag{6.12}$$

by Assumption B2. The bound in Equation 6.11 will therefore tend to 0 as $n \to \infty$, and Equation 6.5 will be *a fortiori* true.

To demonstrate the validity of Inequality 6.10 is the main burden of the proof. By Equation 6.6

$$\begin{aligned}\mathbf{T}_k + \mathbf{T}_k' &= \frac{1}{\Delta_k} \sum_{j \in J_k} (\mathbf{a}_j\mathbf{h}_j' + \mathbf{h}_j\mathbf{a}_j')\\ &= \frac{1}{\Delta_k} \sum_{j \in J_k} r_j(\mathbf{v}_j\mathbf{u}_j' + \mathbf{u}_j\mathbf{v}_j'),\end{aligned} \tag{6.13}$$

where

$$\mathbf{v}_j = \mathbf{a}_j/\|\mathbf{a}_j\|, \qquad \mathbf{u}_j = \mathbf{h}_j/\|\mathbf{h}_j\|, \qquad r_j = \|\mathbf{a}_j\|\,\|\mathbf{h}_j\|.$$

The unit vector $\mathbf{v}_j$ can be decomposed into its components in the direction of $\mathbf{u}_j$ and orthogonal to it:

$$\mathbf{v}_j = \hat{\alpha}_j\mathbf{u}_j + \sqrt{1 - \hat{\alpha}_j^2}\,\hat{\mathbf{u}}_j, \tag{6.14}$$

where

$$\hat{\alpha}_j = \mathbf{u}_j'\mathbf{v}_j, \qquad \mathbf{u}_j'\hat{\mathbf{u}}_j = 0.$$

Here $\hat{\alpha}_j$ is the cosine of the angle between $\mathbf{a}_j$ and $\mathbf{h}_j$ which, by Assumption B5, is positive for all $j \in J_k$ and all large enough k. We assume, hereafter, that the index k is so restricted. Since

$$\lambda_{\min}(\mathbf{T}_k + \mathbf{T}_k') = \min_{\|\mathbf{x}\|=1} \mathbf{x}'(\mathbf{T}_k + \mathbf{T}_k')\mathbf{x},$$

Equations 6.13 and 6.14 yield

$$\begin{aligned}
\lambda_{\min}(\mathbf{T}_k + \mathbf{T}_k') &= \min_{\|\mathbf{x}\|=1} \frac{2}{\Delta_k} \sum_{j \in J_k} r_j(\mathbf{x}'\mathbf{v}_j)(\mathbf{x}'\mathbf{u}_j) \\
&= \min_{\|\mathbf{x}\|=1} \frac{2}{\Delta_k} \sum_{j \in J_k} r_j[\hat{\alpha}_j(\mathbf{x}'\mathbf{u}_j)^2 + \sqrt{1 - \hat{\alpha}_j^2}\,(\mathbf{x}'\hat{\mathbf{u}}_j)(\mathbf{x}'\mathbf{u}_j)] \\
&\geq \min_{\|\mathbf{x}\|=1} \frac{2}{\Delta_k} \Big[\beta_k\alpha_k \sum_{j \in J_k} (\mathbf{x}'\mathbf{u}_j)^2 \\
&\qquad - \gamma_k\sqrt{1 - \alpha_k^2} \sum_{j \in J_k} |\mathbf{x}'\hat{\mathbf{u}}_j|\,|\mathbf{x}'\mathbf{u}_j|\Big], \qquad (6.15)
\end{aligned}$$

where

$$\alpha_k = \min_{j \in J_k} \hat{\alpha}_j, \qquad \beta_k = \min_{j \in J_k} r_j, \qquad \gamma_k = \max_{j \in J_k} r_j.$$

If $\mathbf{u}_j$ and $\hat{\mathbf{u}}_j$ are imbedded in an orthonormal basis for p-space, it follows from the Fourier representation for $\mathbf{x}$ in this coordinate system that

$$1 = \|\mathbf{x}\|^2 \geq (\mathbf{x}'\mathbf{u}_j)^2 + (\mathbf{x}'\hat{\mathbf{u}}_j)^2$$

with equality when $p = 2$. Thus,

$$|\mathbf{x}'\hat{\mathbf{u}}_j| \leq \sqrt{1 - (\mathbf{x}'\mathbf{u}_j)^2}.$$

If we set

$$\xi_j(\mathbf{x}) = |\mathbf{x}'\mathbf{u}_j|, \qquad (6.16)$$

this combines with Equation 6.15 to give

$$\lambda_{\min}(\mathbf{T}_k + \mathbf{T}_k') \geq \min_{\|\mathbf{x}\|=1} \frac{2}{\Delta_k}\Big[\beta_k\alpha_k \sum_{j \in J_k} \xi_j^2(\mathbf{x}) - \gamma_k\sqrt{1 - \alpha_k^2} \sum_{j \in J_k} \xi_j(\mathbf{x})\sqrt{1 - \xi_j^2(\mathbf{x})}\Big]. \qquad (6.17)$$

We successively decrease the lower bound in Equation 6.17 by taking the minimum over larger $\mathbf{x}$-sets in p-space. We have

$$\min_{\|\mathbf{x}\|=1} \sum_{j \in J_k} \xi_j^2(\mathbf{x}) = \min_{\|\mathbf{x}\|=1} \mathbf{x}'\Big(\sum_{j \in J_k} \mathbf{u}_j\mathbf{u}_j'\Big)\mathbf{x} = \lambda_k,$$

where, since $\mathbf{u}_j$ is the normalized $\mathbf{h}_j$,

$$\lambda_k = \lambda_{\min}\Big(\sum_{j\in J_k} \mathbf{h}_j\mathbf{h}_j'/\|\mathbf{h}_j\|^2\Big). \tag{6.18}$$

Thus, the set of all unit length vectors $\mathbf{x}$ is a subset of those for which

$$\sum_{j\in J_k} \xi_j^2(\mathbf{x}) \geq \lambda_k.$$

In turn, the set of all real numbers ξ_j in the unit interval which satisfy

$$\sum_{j\in J_k} \xi_j^2 \geq \lambda_k$$

contains the set of those of the form of Equation 6.16 which satisfy the inequality. Consequently, the lower bound in Equation 6.17 can be weakened to

$$\lambda_{\min}(\mathbf{T}_k + \mathbf{T}_k') \geq \min_{\lambda_k \leq \sum_{j\in J_k} \xi_j^2 \leq p_k} \Big[A_k \sum_{j\in J_k} \xi_j^2 - B_k \sum_{j\in J_k} \xi_j\sqrt{1-\xi_j^2}\Big],$$

where we have set

$$A_k = \frac{2\beta_k\alpha_k}{\Delta_k}, \qquad B_k = \frac{2\gamma_k\sqrt{1-\alpha_k^2}}{\Delta_k}. \tag{6.19}$$

After applying the Schwarz Inequality to the second term on the right-hand side and setting

$$z = \sqrt{\frac{1}{p_k}\sum_{j\in J_k} \xi_j^2}, \qquad g_k(z) = z^2 - \left(\frac{B_k}{A_k}\right) z\sqrt{1-z^2},$$

we obtain

$$\lambda_{\min}(\mathbf{T}_k + \mathbf{T}_k') \geq p_k A_k \min_{\sqrt{\lambda_k/p_k} \leq z \leq 1} g_k(z). \tag{6.20}$$

Inequality 6.10 will thus follow if the lower bound in Inequality 6.20 has a strictly positive limit inferior as $k \to \infty$. We now complete the proof by showing, as a consequence of Assumptions B3 through B5, that this is indeed the case.

In the original notation, the numerator quantities in Equation 6.19 are

$$\alpha_k = \min_{j\in J_k} \frac{\mathbf{a}_j'\mathbf{h}_j}{\|\mathbf{a}_j\|\ \|\mathbf{h}_j\|} \qquad \beta_k = \min_{j\in J_k} \|\mathbf{a}_j\|\ \|\mathbf{h}_j\|, \qquad \gamma_k = \max_{j\in J_k} \|\mathbf{a}_j\|\ \|\mathbf{h}_j\|,$$

while the common denominator Δ_k is given by Equation 6.7. Since $\Delta_k^2 \leq p_k\gamma_k^2$, we have

$$\liminf_k p_k A_k \geq \liminf_k 2\sqrt{p_k}\,\frac{\beta_k}{\gamma_k}\,\alpha_k > 2\sqrt{p}\,\frac{\alpha}{\rho} > 0 \tag{6.21}$$

according to Assumptions B4 and B5. From these two assumptions, it also follows that

$$\limsup_k \frac{B_k}{A_k} = \frac{\tau}{\sqrt{1 - \tau^2}} - 2\delta$$

for some $\delta > 0$. For any such number, we see that

$$\tau - \frac{\tau - \delta\sqrt{1 - \tau^2}}{\sqrt{(1 - \tau^2) + [\tau - \delta\sqrt{1 - \tau^2}]^2}} = 2\varepsilon > 0, \tag{6.22}$$

because the left-hand side increases steadily from zero when viewed as a function of δ. Using Equation 6.18 and Assumption B3, we now fix $k(\delta)$ so large that

$$\frac{B_k}{A_k} \leq \frac{\tau}{\sqrt{1 - \tau^2}} - \delta \qquad \text{and} \qquad \sqrt{\frac{\lambda_k}{p_k}} \geq \liminf_k \sqrt{\frac{\lambda_k}{p_k}} - \varepsilon = \tau - \varepsilon$$

hold simultaneously for $k \geq k(\delta)$. For all such indices, we can, therefore, write

$$\min_{\sqrt{\lambda_k/p_k} \leq z \leq 1} g_k(z) \geq \min_{\tau - \varepsilon \leq z \leq 1} g(z), \tag{6.23}$$

where

$$g(z) = z^2 - \left(\frac{\tau}{\sqrt{1 - \tau^2}} - \delta\right) z\sqrt{1 - z^2}.$$

This function is strictly convex on [0, 1], as can be easily seen from an examination of $g(\sin\theta)$ over $0 \leq \theta \leq \pi/2$. It has roots at $z = 0$ and at

$$z_0 = \frac{\tau - \delta\sqrt{1 - \tau^2}}{\sqrt{(1 - \tau^2) + [\tau - \delta\sqrt{1 - \tau^2}]^2}} = \tau - 2\varepsilon,$$

the last by the definition of Equation 6.22. Therefore, $g(\cdot)$ must be strictly positive over $[\tau - \varepsilon, 1]$, because $z_0 < \tau - \varepsilon$. This, together with Equations 6.23 and 6.21, implies the desired conclusion for Equation 6.20. Q.E.D.

Let us now return to the sequence of estimates (Equation 6.1), and focus our attention on the resulting difference equation (Equation 6.3). We allow the gain vector $\mathbf{a}_j$ to depend on the first j iterates, so that the leading product is written

$$\mathbf{P}_n(\mathbf{t}_1, \cdots, \mathbf{t}_n) = \prod_{j=1}^{n} [\mathbf{I} - \mathbf{a}_j(\mathbf{t}_1, \cdots, \mathbf{t}_j)\mathbf{h}_j'(\mathbf{t}_j)].$$

In keeping with the notation of Theorem 6.1, we have set

$$\dot{\mathbf{F}}_j(\mathbf{u}_j) = \mathbf{h}_j(\mathbf{t}_j),$$

which is indeed the case because $\mathbf{u}_j$ depends on the iterates only through the value of $\mathbf{t}_j$. It should be clear that the above sequence of matrix-valued random variables $\{\mathbf{P}_n\}$ will (surely) tend to the zero matrix as $n \to \infty$, if we require that the vectors $\mathbf{a}_n(\mathbf{x}_1, \cdots, \mathbf{x}_n)$ and $\dot{\mathbf{F}}_n(\mathbf{y})$ satisfy Assumptions B1 through B5 of Theorem 6.1 uniformly in all vector arguments $\mathbf{x}_1, \cdots, \mathbf{x}_n$ and $\mathbf{y}$. Such are, in fact, the first five assumptions of the following theorem. The sixth takes care of the additional term in Equation 6.3 arising from the stochastic residuals $W_1, W_2, \cdots$ and ensures mean-square convergence of $\|\mathbf{t}_n - \boldsymbol{\theta}\|$ to zero.

6.3 Theorems Concerning Mean-Square and Probability-One Convergence for General Gains and Assumptions (C1 Through C6′ and D1 Through D5)

THEOREM 6.2

Let $\{\mathbf{Y}_n : n = 1, 2, \cdots\}$ be a real-valued stochastic process of the form $Y_n = F_n(\boldsymbol{\theta}) + W_n$, where $F_n(\cdot)$ is known up to the p-dimensional parameter $\boldsymbol{\theta}$, and $W_1, W_2, \cdots$ have uniformly bounded variances. For each n, let $\mathbf{a}_n(\cdot)$ be a Borel measurable mapping of the product space $\times_1^n R^p$ into R^p (Euclidean p-space), and let

$$\mathbf{t}_{n+1} = \mathbf{t}_n + \mathbf{a}_n(\mathbf{t}_1, \cdots, \mathbf{t}_n)[Y_n - F_n(\mathbf{t}_n)] \qquad (n = 1, 2, \cdots; \mathbf{t}_1 \text{ arbitrary}).$$

Denote the gradient vector of F_n by $\dot{\mathbf{F}}_n$ and suppose the following assumptions hold:

C1. $\lim_n \sup_{\mathbf{x}_1, \cdots, \mathbf{x}_n, \mathbf{y}} \|\mathbf{a}_n(\mathbf{x}_1, \cdots, \mathbf{x}_n)\| \, \|\dot{\mathbf{F}}_n(\mathbf{y})\| = 0.$

C2. $\sum_n \inf_{\mathbf{x}_1, \cdots, \mathbf{x}_n, \mathbf{y}} \|\mathbf{a}_n(\mathbf{x}_1, \cdots, \mathbf{x}_n)\| \, \|\dot{\mathbf{F}}_n(\mathbf{y})\| = \infty.$

C3. There exists a sequence of integers $1 = \nu_1 < \nu_2 < \nu_3 \cdots$ such that, with $p_k = \nu_{k+1} - \nu_k$,

$$p \le p_k \le q < \infty \qquad (k = 1, 2, \cdots),$$

and

$$\liminf_k \frac{1}{p_k} \inf_{\mathbf{y}_{\nu_k}, \cdots, \mathbf{y}_{\nu_{k+1}-1}} \lambda_{\min}\left(\sum_{j \in J_k} \frac{\dot{\mathbf{F}}_j(\mathbf{y}_j)\dot{\mathbf{F}}_j'(\mathbf{y}_j)}{\|\dot{\mathbf{F}}_j(\mathbf{y}_j)\|^2}\right) = \tau^2 > 0,$$

where

$$J_k = \{\nu_k, \nu_k + 1, \cdots, \nu_{k+1} - 1\}.$$

C4. $$\limsup_k \sup_{\mathbf{x}_1,\cdots,\mathbf{x}_{\nu_{k+1}-1},\mathbf{y}_{\nu_k},\cdots,\mathbf{y}_{\nu_{k+1}-1}} \frac{\max_{j\in J_k} \|\mathbf{a}_j(\mathbf{x}_1,\cdots,\mathbf{x}_j)\| \, \|\dot{\mathbf{F}}_j(\mathbf{y}_j)\|}{\min_{j\in J_k} \|\mathbf{a}_j(\mathbf{x}_1,\cdots,\mathbf{x}_j)\| \, \|\dot{\mathbf{F}}_j(\mathbf{y}_j)\|} = \rho < \infty.$$

C5. $$\liminf_n \inf_{\mathbf{x}_1,\cdots,\mathbf{x}_n,\mathbf{y}} \frac{\mathbf{a}_n'(\mathbf{x}_1,\cdots,\mathbf{x}_n)\dot{\mathbf{F}}_n(\mathbf{y})}{\|\mathbf{a}_n(\mathbf{x}_1,\cdots,\mathbf{x}_n)\| \, \|\dot{\mathbf{F}}_n(\mathbf{y})\|} > \alpha,$$

where

$$\alpha = \sqrt{\frac{1-\tau^2}{1-\tau^2+(\tau/\rho)^2}}.$$

Then $\mathscr{E}\|\mathbf{t}_n - \boldsymbol{\theta}\|^2 \to 0$ as $n \to \infty$ if either

C6. $\sum_n \sup_{\mathbf{x}_1,\cdots,\mathbf{x}_n} \|\mathbf{a}_n(\mathbf{x}_1,\cdots,\mathbf{x}_n)\| < \infty$ or

C6′. $\{W_n\}$ is a zero-mean independent process and

$$\sum_n \sup_{\mathbf{x}_1,\cdots,\mathbf{x}_n} \|\mathbf{a}_n(\mathbf{x}_1,\cdots,\mathbf{x}_n)\|^2 < \infty.$$

Proof. The argument consists of three main parts. As in the proof of Theorem 6.1, given any n, let $K = K(n)$ be the index such that

$$\nu_K \le n \le \nu_{K+1} - 1.$$

We first show, with a minimum of effort, that

$$\mathscr{E}\|\mathbf{t}_{n+1} - \boldsymbol{\theta}\|^2 \le M_1\mathscr{E}\|\mathbf{t}_{\nu_K} - \boldsymbol{\theta}\|^2 + M_2\mathscr{E}^{1/2}\|\mathbf{t}_{\nu_K} - \boldsymbol{\theta}\|^2 + M_3 A_K^2, \tag{6.24}$$

where $A_K \to 0$ as $K \to \infty$ with n. (M's, with or without affixes, will stand for finite constants which will be reused without regard to meaning.) It is more than sufficient, therefore, to have

$$e_k^2 = \mathscr{E}\|\mathbf{t}_{\nu_k} - \boldsymbol{\theta}\|^2 \to 0 \tag{6.25}$$

as $k \to \infty$ over the integers. This is immediate, if we can prove that

$$e_{k+1}^2 \le (1 - b\Delta_k)e_k^2 + M_4 B_k \tag{6.26}$$

holds for all large enough k, say $k \ge N$, and some sequences having the properties

$$b\Delta_k > 0, \qquad \lim \Delta_k = 0, \qquad \Sigma\Delta_k = \infty, \qquad \Sigma B_k < \infty. \tag{6.27}$$

Indeed, after iterating Equation 6.26 back to N, we obtain

$$e_{k+1}^2 \le \prod_{j=N}^{k} (1 - b\Delta_j)e_N^2 + M_4 \sum_{j=N}^{k} \prod_{i=j+1}^{k} (1 - b\Delta_i)B_j.$$

It follows from Equation 6.27 and Lemma 2 that this upper bound goes to zero as k tends to infinity. The sought-after conclusion will thus be at hand.

The second and third parts of the proof establish Equations 6.26 and 6.27 under Assumptions C6 and C6′, respectively. In the former case, the argument is relatively straightforward. Under Assumption C6′, however, the details are a bit more complicated, but we are finally able to use the independence to establish the desired inequality with some (other) sequences which obey Equation 6.27.

Proof of Equation 6.24. Iterate Equation 6.2 back from $n + 1$ to ν_K, where $K = K(n)$ is as before. We obtain

$$\mathbf{t}_{n+1} - \boldsymbol{\theta} = \prod_{j=\nu_K}^{n} (\mathbf{I} - \mathbf{a}_j\mathbf{h}_j')(\mathbf{t}_{\nu_K} - \boldsymbol{\theta}) + \sum_{j=\nu_K}^{n} \prod_{i=j+1}^{n} (\mathbf{I} - \mathbf{a}_i\mathbf{h}_i')\mathbf{a}_j W_j, \tag{6.28}$$

where it will be necessary to remember that $\mathbf{a}_j$ and $\mathbf{h}_j$ are now vector-valued random variables:

$$\mathbf{a}_j = \mathbf{a}_j(\mathbf{t}_1, \cdots, \mathbf{t}_j), \qquad \mathbf{h}_j = \dot{\mathbf{F}}_j(\mathbf{u}_j). \tag{6.29}$$

We "square" both sides of Equation 6.28, take expectations, and then bound from above (in the obvious way) the two squared norms and the inner product. The result is

$$\begin{aligned}
&\mathscr{E}\|\mathbf{t}_{n+1} - \boldsymbol{\theta}\|^2 \\
&\quad\le \mathscr{E}\left[\prod_{j=\nu_K}^{n} \|\mathbf{I} - \mathbf{a}_j\mathbf{h}_j'\|\right]^2 \|\mathbf{t}_{\nu_K} - \boldsymbol{\theta}\|^2 \\
&\quad+ 2\mathscr{E}\left\{\|\mathbf{t}_{\nu_K} - \boldsymbol{\theta}\| \prod_{j=\nu_K}^{n} \|\mathbf{I} - \mathbf{a}_j\mathbf{h}_j'\| \sum_{j=\nu_K}^{n} \prod_{i=j+1}^{n} \|\mathbf{I} - \mathbf{a}_i\mathbf{h}_i'\| \, \|\mathbf{a}_j\| \, |W_j|\right\} \\
&\quad+ \mathscr{E}\left[\sum_{j=\nu_K}^{n} \prod_{i=j+1}^{n} \|\mathbf{I} - \mathbf{a}_i\mathbf{h}_i\| \, \|\mathbf{a}_j\| \, |W_j|\right]^2.
\end{aligned} \tag{6.30}$$

From Assumption C1 it follows that there is a real number M such that $\sup_n \|\mathbf{I} - \mathbf{a}_n\mathbf{h}_n'\| < M$; therefore, by Assumption C3, we have

$$\prod_{j=m}^{n} \|\mathbf{I} - \mathbf{a}_j\mathbf{h}_j'\| < M^q \tag{6.31}$$

for all $\nu_k \le m \le n \le \nu_{k+1} - 1$ and all $k \ge 1$. (Unqualified deterministic bounds on random variables are to be read as sure ones.) Under either Assumptions C6 or C6′, we have

$$\max_{j \in J_k} \|\mathbf{a}_j\| \le \max_{j \in J_k} \sup_{\mathbf{x}_1, \cdots, \mathbf{x}_j} \|\mathbf{a}_j(\mathbf{x}_1, \cdots, \mathbf{x}_j)\| = A_k \to 0 \tag{6.32}$$

as $k \to \infty$. If we apply these results with $k = K(n)$ to Equation 6.30 and then use the Schwarz Inequality, we get

$$M^{-2q}\mathscr{E}\|\mathbf{t}_{n+1} - \boldsymbol{\theta}\|^2 \le \mathscr{E}\|\mathbf{t}_{\nu_K} - \boldsymbol{\theta}\|^2 + 2A_K\mathscr{E}^{1/2}\|\mathbf{t}_{\nu_K} - \boldsymbol{\theta}\|^2\mathscr{E}^{1/2}\Big(\sum_{j\in J_K}|W_j|\Big)^2 + A_K{}^2\mathscr{E}\Big(\sum_{j\in J_K}|W_j|\Big)^2.$$

From Equation 6.32 and this, Equation 6.24 follows, because J_k contains no more than q indices for any k, and $\sup_n \mathscr{E}W_n{}^2$ is finite by hypothesis.

Proof of Equations 6.26 and 6.27 under Assumption C6. We return to Equation 6.28, set $n = \nu_{K+1} - 1$, and then replace K by an arbitrary integer k. After again "squaring," we have

$$\|\mathbf{t}_{\nu_{k+1}} - \boldsymbol{\theta}\|^2 = \|\mathbf{Q}_k\|^2\|\mathbf{t}_{\nu_k} - \boldsymbol{\theta}\|^2 + 2(\mathbf{t}_{\nu_k} - \boldsymbol{\theta})'\mathbf{Q}_k' \sum_{j\in J_k}\prod_{i=j+1}^{\nu_{k+1}-1}(\mathbf{I} - \mathbf{a}_i\mathbf{h}_i')\mathbf{a}_j W_j + \Big\|\sum_{j\in J_k}\prod_{i=j+1}^{\nu_{k+1}-1}(\mathbf{I} - \mathbf{a}_i\mathbf{h}_i')\mathbf{a}_j W_j\Big\|^2, \tag{6.33}$$

where, in contrast to Equation 6.6,

$$\mathbf{Q}_k = \prod_{j\in J_k}(\mathbf{I} - \mathbf{a}_j\mathbf{h}_j') = \mathbf{Q}_k(\mathbf{t}_1, \cdots, \mathbf{t}_{\nu_{k+1}-1}) \qquad (k = 1, 2, \cdots) \tag{6.34}$$

is stochastic. The deterministic quantity to be used here in place of Equation 6.7 is

$$\Delta_k = \Big(\sum_{j\in J_k}\inf_{\mathbf{x}_1,\cdots,\mathbf{x}_j,\mathbf{y}}\|\mathbf{a}_j(\mathbf{x}_1, \cdots, \mathbf{x}_j)\|^2\|\dot{\mathbf{F}}_j(\mathbf{y})\|^2\Big)^{1/2}. \tag{6.35}$$

Because of Assumptions C1 and C3, we have

$$\lim_k \Delta_k = 0. \tag{6.36}$$

We formally define $\mathbf{T}_k$ the way we did in Equation 6.6, but with the summands given by Equation 6.29 and Δ_k by Equation 6.35. Using Assumption C4 in addition to C1, we see that Equation 6.9 remains true for the matrix $\mathbf{Q}_k$ of Equation 6.34. Furthermore, by virtue of the uniform nature of Assumptions C3 through C5, the same (long) type of argument which led to Equation 6.10 proves, for the present situation, that

$$\|\mathbf{Q}_k\|^2 \le (1 - c\Delta_k)^2 \qquad (k \ge N) \tag{6.37}$$

holds for some (deterministic) $c > 0$ and $N < \infty$. We now apply the

Schwarz Inequality to the second term on the right-hand side of Equation 6.33 and then majorize the bound using Equations 6.31, 6.32, and 6.37. After taking expectations, we obtain for $k \geq N$, in the notation of Equation 6.25,

$$\begin{aligned} e_{k+1}^2 &\leq (1 - c\Delta_k)^2 e_k^2 + 2M^{2q} A_k e_k \mathscr{E}^{1/2}\Big(\sum_{j \in J_k} |W_j|\Big)^2 \\ &\quad + M^{2q} A_k^2 \mathscr{E}\Big(\sum_{j \in J_k} |W_j|\Big)^2 \\ &\leq (1 - c\Delta_k)^2 e_k^2 + M_1 A_k e_k + M_2 A_k^2 \\ &\leq (1 - c\Delta_k)^2 e_k^2 + M_3 A_k (1 + e_k), \end{aligned} \tag{6.38}$$

because second-order noise moments are uniformly bounded. For A_k defined in Equation 6.32, we have, after making use of Assumption C6,

$$\Sigma A_k < \infty, \tag{6.39}$$

because the maximum value of a number of terms is certainly smaller than their sum. With Equations 6.36 and 6.39 we satisfy the hypotheses of Lemma 3, and by its conclusion have

$$\sup_k e_k^2 < \infty.$$

Hence, from Equation 6.38 there results

$$e_{k+1}^2 \leq \left(1 - \frac{3}{2} c\Delta_k\right) e_k^2 + M_4 A_k \tag{6.40}$$

(say) for all large enough k. It remains to be seen (for the same reason that Equation 6.12 followed from Assumption B2) that Assumption C2 implies

$$\Sigma \Delta_k = \infty \tag{6.41}$$

for the sequence of Equation 6.35.

Proof of Equations 6.26 and 6.27 under Assumption C6′. **When Equation 6.39 fails to hold, we need a tighter bound on the expectation of the cross term in Equation 6.33 than the one used in Equation 6.38. Specifically, we will show that**

$$L = \left| \mathscr{E}(\mathbf{t}_{\nu_k} - \boldsymbol{\theta})' \mathbf{Q}_k' \sum_{j \in J_k} \prod_{i=j+1}^{\nu_{k+1}-1} (\mathbf{I} - \mathbf{a}_i \mathbf{h}_i') \mathbf{a}_j W_j \right| \leq c\Delta_k e_k^2 + M_1 \Delta_k A_k^2 \tag{6.42}$$

no matter what $c > 0$ in Equation 6.37. All results of the previous

paragraph, with the exception of 6.39, were derived from Assumptions C1 through C5; in particular, the balance of the second bound on e_{k+1}^2 in Equation 6.38 remains true as written. Thus, given the validity of Equation 6.42, we will have, because $\Delta_k \to 0$,

$$\begin{aligned} e_{k+1}^2 &\leq (1 - c\Delta_k)^2 e_k{}^2 + c\Delta_k e_k{}^2 + M_1 \Delta_k A_k{}^2 + M_2 A_k{}^2 \\ &\leq (1 - \tfrac{1}{2}c\Delta_k)^2 e_k{}^2 + M_3 A_k{}^2 \end{aligned} \tag{6.43}$$

for all large enough k. Just as Assumption C6 led to Equation 6.39, we see that Assumption C6′ implies that

$$\Sigma A_k{}^2 < \infty. \tag{6.44}$$

Equation 6.43 is the desired inequality of Equation 6.26, while Equations 6.36, 6.41, and 6.44 are collectively the statement of Equation 6.27.

It remains, therefore, to establish Equation 6.42. We begin by carrying out the matrix multiplication called for. Using the definition of Equation 6.34, we find that

$$\mathbf{Q}_k' \prod_{i=j+1}^{\nu_{k+1}-1} (\mathbf{I} - \mathbf{a}_i \mathbf{h}_i') = \left(\mathbf{I} + \sum_{i=\nu_k}^{j} \mathbf{a}_i \mathbf{h}_i'\right) - \Delta_k(\mathbf{T}_k + \mathbf{T}_k') + \Delta_k{}^2 \mathbf{E}_{jk},$$

where, as earlier, Δ_k is given by Equation 6.35 and

$$\|\mathbf{T}_k\| \leq \frac{1}{\Delta_k} \sum_{j \in J_k} \|\mathbf{a}_j \mathbf{h}_j\| < M_1,$$

in view of Assumption C4. After much manipulation, it turns out that the norm of the matrix $\mathbf{E}_{jk}$ is also uniformly bounded:

$$\sup_{k \geq 1} \sup_{j \in J_k} \|\mathbf{E}_{jk}\| < M_2.$$

Thus, the left-hand side of Equation 6.42 is equal to

$$\begin{aligned} L = \Big|\mathscr{E}(\mathbf{t}_{\nu_k} - \boldsymbol{\theta})' &\sum_{j \in J_k} \left(\mathbf{I} + \sum_{i=\nu_k}^{j} \mathbf{a}_i \mathbf{h}_i'\right) \mathbf{a}_j W_j \\ &- \mathscr{E}(\mathbf{t}_{\nu_k} - \boldsymbol{\theta})' \Delta_k (\mathbf{T}_k + \mathbf{T}_k') \sum_{j \in J_k} \mathbf{a}_j W_j \\ &+ \mathscr{E}(\mathbf{t}_{\nu_k} - \boldsymbol{\theta})' \Delta_k{}^2 \sum_{j \in J_k} \mathbf{E}_{jk} \mathbf{a}_j W_j\Big| = |\mathrm{I} + \mathrm{II} + \mathrm{III}|. \end{aligned} \tag{6.45}$$

Consider the first term. Since $\mathbf{a}_i$ and $\mathbf{h}_i$ depend on the estimates up through time i, and hence on the observational errors up through time $i - 1$, W_j is independent of

$$\mathbf{t}_{\nu_k}, \qquad \sum_{i=\nu_k}^{j} \mathbf{a}_i \mathbf{h}_i', \qquad \text{and} \qquad \mathbf{a}_j$$

for all $j \in J_k$. Consequently, we have

$$\mathscr{E}\left\{(\mathbf{t}_{\nu_k} - \boldsymbol{\theta})'\left[\mathbf{I} + \sum_{i=\nu_k}^{j} \mathbf{a}_i\mathbf{h}_i'\right]\mathbf{a}_j W_j \middle| W_1, \cdots, W_{j-1}\right\}$$

$$= (\mathbf{t}_{\nu_k} - \boldsymbol{\theta}')\left[\mathbf{I} + \sum_{i=\nu_k}^{j} \mathbf{a}_i\mathbf{h}_i'\right]\mathbf{a}_j\mathscr{E}\{W_j | W_1, \cdots, W_{j-1}\} = 0$$

for all $j \in J_k$ and, hence, each of the p_k unconditional expectations must vanish, giving

$$|\mathrm{I}| = 0. \tag{6.46}$$

For the second term, in view of Equation 6.32, we have

$$|\mathrm{II}| \le \mathscr{E} \|\mathbf{t}_{\nu_k} - \boldsymbol{\theta}\| \, \Delta_k \, \|\mathbf{T}_k + \mathbf{T}_k'\| \sum_{j \in J_k} \|\mathbf{a}_j\| \, |W_j| \le M_3\Delta_k A_k e_k, \tag{6.47}$$

where M_3 involves M_1 (with the last given meaning) and the uniform bound on residual variances. Similarly, we have

$$|\mathrm{III}| \le \mathscr{E} \|\mathbf{t}_{\nu_k} - \boldsymbol{\theta}\| \, \Delta_k^2 \sum_{j \in J_k} \|\mathbf{E}_{jk}\| \, \|\mathbf{a}_j\| \, |W_j| \le M_4\Delta_k^2 A_k e_k, \tag{6.48}$$

where M_4 involves M_2, and so on. Since $\Delta_k \to 0$ as $k \to \infty$, Equations 6.45 through 6.48 combine to give

$$L \le M_5\Delta_k A_k e_k \equiv [c\Delta_k e_k^2]^{1/2} \left[\frac{M_5^2}{c} \Delta_k A_k^2\right]^{1/2},$$

where the identity is in the number $c > 0$ appearing in Equation 6.37. Calling these two square roots a and b, respectively, we obtain

$$L \le c\Delta_k e_k^2 + M_6\Delta_k A_k^2$$

from the generally valid inequality $ab \le a^2 + b^2$. This is the assertion of Equation 6.42. Q.E.D.

The proof of almost sure convergence can be executed under Assumptions C1 through C5 and either Assumption C6 or C6′. Actually, Assumptions C1 through C5 can be weakened somewhat, and it is under such a set of weaker conditions that we prove the following.

THEOREM 6.3

In the notation and under the assumptions of Theorem 6.2, $\mathbf{t}_n$ converges to $\boldsymbol{\theta}$ with probability one as $n \to \infty$. This remains true if Assumptions C1 through C5 are replaced by the following conditions: For every sequence of p-vectors $\mathbf{x}_1, \mathbf{x}_2, \cdots$ and $\mathbf{y}_1, \mathbf{y}_2, \cdots$,

D1. $\lim_n \|\mathbf{a}_n(\mathbf{x}_1, \cdots, \mathbf{x}_n)\| \, \|\dot{\mathbf{F}}_n(\mathbf{y}_n)\| = 0.$

D2. $\sum_n \|\mathbf{a}_n(\mathbf{x}_1, \cdots, \mathbf{x}_n)\| \, \|\dot{\mathbf{F}}_n(\mathbf{y}_n)\| = \infty.$

D3. There exists a sequence of integers $1 = \nu_1 < \nu_2 < \nu_3 \cdots$ such that, with $p_k = \nu_{k+1} - \nu_k$,

$$p \le p_k \le q < \infty \qquad (k = 1, 2, \cdots)$$

and

$$\liminf_n \frac{1}{p_k} \lambda_{\min}\left(\sum_{j \in J_k} \frac{\dot{\mathbf{F}}_j(\mathbf{y}_j)\dot{\mathbf{F}}_j'(\mathbf{y}_j)}{\|\dot{\mathbf{F}}_j(\mathbf{y}_j)\|^2}\right) = \tau^2 > 0,$$

where

$$J_k = \{\nu_k, \nu_k + 1, \cdots, \nu_{k+1} - 1\}.$$

D4. $$\limsup_k \frac{\max_{j \in J_k} \|\mathbf{a}_j(\mathbf{x}_1, \cdots, \mathbf{x}_j)\| \, \|\dot{\mathbf{F}}_j(\mathbf{y}_j)\|}{\min_{j \in J_k} \|\mathbf{a}_j(\mathbf{x}_1, \cdots, \mathbf{x}_j)\| \, \|\dot{\mathbf{F}}_j(\mathbf{y}_j)\|} = \rho < \infty.$$

D5. $$\liminf_n \frac{\mathbf{a}_n'(\mathbf{x}_1, \cdots, \mathbf{x}_n)\dot{\mathbf{F}}_n(\mathbf{y}_n)}{\|\mathbf{a}_n(\mathbf{x}_1, \cdots, \mathbf{x}_n)\| \, \|\dot{\mathbf{F}}_n(\mathbf{y}_n)\|} > \alpha = \sqrt{\frac{1 - \tau^2}{1 - \tau^2 + (\tau/\rho)^2}}.$$

The ν_k's, τ^2, and ρ can depend on the sequences $\{\mathbf{x}_n\}$ and $\{\mathbf{y}_n\}$.

Proof. Let $\mathbf{t}_1, \mathbf{t}_2, \mathbf{t}_3, \cdots$ be a realization of the random-variable sequence recursively defined in the statement of Theorem 6.2. From Equation 6.28, we have

$$\|\mathbf{t}_{n+1} - \boldsymbol{\theta}\| \le \prod_{j=\nu_K}^{n} \|\mathbf{I} - \mathbf{a}_j\mathbf{h}_j'\| \, \|\mathbf{t}_{\nu_K} - \boldsymbol{\theta}\| + \sum_{j=\nu_K}^{n} \prod_{i=j+1}^{n} \|\mathbf{I} - \mathbf{a}_i\mathbf{h}_i'\| \, \|\mathbf{a}_j W_j\|, \quad (6.48)$$

where $K = K(n)$ is the (now, possibly random) integer defined at the outset of the proof of Theorem 6.2. The random vectors $\mathbf{a}_j$ and $\mathbf{h}_j$ in Equation 6.29 are seen to satisfy Assumptions D1 through D5 with probability one, when we set $\mathbf{x}_j = \mathbf{t}_j$ and $\mathbf{y}_j = \mathbf{u}_j = \mathbf{u}_j(\mathbf{t}_j)$. Let

$$\mathbf{s}_{n+1} = \sum_{k=1}^{n} \mathbf{a}_j W_j. \quad (6.49)$$

The same two arguments used in the final paragraph of the proof of Theorem 2.1, to show that Equation 2.13 follows from Condition 4 or 5, apply here to show (component-wise) that

$$\mathbf{s}_n \xrightarrow{\text{a.s.}} \mathbf{s} \quad (6.50)$$

follows from Assumptions C6 or C6′ for some finite vector-valued random variable $\mathbf{s}$. We will not actually use Equation 6.50 until later on, and for the present we only make note of the fact that it implies

$$\|\mathbf{a}_n(\mathbf{t}_1, \cdots, \mathbf{t}_n)W_n\| \overset{\text{a.s.}}{\to} 0 \tag{6.51}$$

as $n \to \infty$. Under Assumption D1, there is a scalar-valued random variable $M = M(\mathbf{t}_1, \mathbf{t}_2, \cdots)$ such that Equation 6.31 holds with probability one. Thus, from Equation 6.48,

$$\|\mathbf{t}_{n+1} - \boldsymbol{\theta}\| \le M^q \|\mathbf{t}_{\nu_K} - \boldsymbol{\theta}\| + qM^q \max_{j \in J_K} \|\mathbf{a}_j W_j\|,$$

and hence, because of Equation 6.51, $\mathbf{t}_n$ will converge a.s. to $\boldsymbol{\theta}$ once we have shown that

$$\|\mathbf{t}_{\nu_m} - \boldsymbol{\theta}\| \overset{\text{a.s.}}{\to} 0 \tag{6.52}$$

as m tends to infinity over the positive integers.

Rather than derive a recursive relationship between the successive deviations in Equation 6.52 (as was previously done for their mean squares, e_m^2), we set $n + 1 = \nu_{m+1}$ in Equation 6.3 and consider the resulting formula for $\mathbf{t}_{\nu_{m+1}} - \boldsymbol{\theta}$ in terms of all past iterates, that is, those with consecutive indices in $J_1, J_2, \cdots, J_m$. (Note the dummy index k in the proof of Theorem 6.2 is here replaced by m.) We first rewrite the sum on the right-hand side of Equation 6.3, with n left arbitrary. We define

$$\mathbf{R}_j = \begin{cases} \prod_{i=j}^{n} (\mathbf{I} - \mathbf{a}_i \mathbf{h}_i') & \text{for } j = 1, 2, \cdots, n, \\ \mathbf{I} & \text{for } j \ge n + 1, \end{cases}$$

where the n-dependence must be remembered. The sum in question is then

$$\begin{aligned} \sum_{j=1}^{n} \mathbf{R}_{j+1}\mathbf{a}_j W_j &= \sum_{j=1}^{n} \mathbf{R}_{j+1}\mathbf{s}_{j+1} - \mathbf{R}_{j+1}\mathbf{s}_j \\ &= \sum_{j=1}^{n} (\mathbf{R}_j \mathbf{s}_j - \mathbf{R}_{j+1}\mathbf{s}_j) + \mathbf{s}_{n+1} \\ &= \sum_{j=1}^{n} [\mathbf{R}_{j+1}(\mathbf{I} - \mathbf{a}_j \mathbf{h}_j)\mathbf{s}_j - \mathbf{R}_{j+1}\mathbf{s}_j] + \mathbf{s}_{n+1} \\ &= -\sum_{j=1}^{n} \mathbf{R}_{j+1}\mathbf{a}_j \mathbf{h}_j' \mathbf{s}_j + \mathbf{s}_{n+1}. \end{aligned}$$

But Lemma 1 tells us that $\sum_{j=1}^{n} \mathbf{R}_{j+1}\mathbf{a}_j\mathbf{h}_j' = \mathbf{I} - \mathbf{R}_1$. We can thereby incorporate the established limit of Equation 6.50 into the result and obtain, with $\mathbf{s}_1 = 0$,

$$\sum_{j=1}^{n} \mathbf{R}_{j+1}\mathbf{a}_j W_j = \mathbf{R}_1\mathbf{s} + (\mathbf{s}_{n+1} - \mathbf{s}) - \sum_{j=1}^{n} \mathbf{R}_{j+1}\mathbf{a}_j\mathbf{h}_j'(\mathbf{s}_j - \mathbf{s}). \quad (6.53)$$

Now, taking $n = \nu_{m+1} - 1$, from Equations 6.3 and 6.53 we obtain

$$\mathbf{t}_{\nu_{m+1}} - \boldsymbol{\theta} = \prod_{k=1}^{m} \mathbf{Q}_k(\mathbf{t}_{\nu_1} - \boldsymbol{\theta} + \mathbf{s}) + (\mathbf{s}_{\nu_{m+1}} - \mathbf{s}) - \sum_{k=1}^{m} \sum_{j\in J_k} \prod_{i=j+1}^{\nu_{m+1}-1} (\mathbf{I} - \mathbf{a}_i\mathbf{h}_i')\mathbf{a}_j\mathbf{h}_j'(\mathbf{s}_j - \mathbf{s}), \quad (6.54)$$

where $\mathbf{Q}_k$ is given by Equation 6.34.

To show that the norm of the leading product matrix goes to 0 a.s. as $m \to \infty$ is an easy matter. We set

$$\Delta_k = \left(\sum_{j\in J_k} \|\mathbf{a}_j\|^2\|\mathbf{h}_j\|^2\right)^{1/2} = \Delta_k(\mathbf{t}_1, \mathbf{t}_2, \cdots, \mathbf{t}_{\nu_{k+1}-1}), \quad (6.55)$$

in contrast to the deterministic definition of Equation 6.35. However, it follows from Assumptions D1 through D3, just as Equations 6.36 and 6.41 did from the uniform version of these assumptions, that

$$\Delta_k \overset{\text{a.s.}}{\to} 0, \qquad \sum_{i=1}^{k} \Delta_i \overset{\text{a.s.}}{\to} \infty \quad (6.56)$$

as $k \to \infty$. Furthermore, Assumptions D3 through D5 imply the existence of a random variable $c > 0$ and an integer-valued random variable $N < \infty$ such that

$$0 \le \|\mathbf{Q}_k\| \le 1 - c\Delta_k \qquad \text{for all} \quad k \ge N \quad (6.57)$$

with probability one. Thus, by Equation 6.56, we have

$$\left\|\prod_{k=1}^{m} \mathbf{Q}_k\right\| \le \left\|\prod_{k=1}^{N-1} \mathbf{Q}_k\right\| \prod_{k=N}^{m} (1 - c\Delta_k) \le C\exp\left\{-c\sum_{k=N}^{n} \Delta_k\right\} \overset{\text{a.s.}}{\to} 0 \quad (6.58)$$

as $n \to \infty$, where C is some positive random variable. It follows that the norm of the first vector on the right-hand side of Equation 6.54 goes to zero as $m \to \infty$. That the second does also is an immediate consequence of Equation 6.50 and the fact that $\nu_m \to \infty$ with m.

It remains to appropriately bound the norm of the third term in Equation 6.54. For any $j \in J_k$, Equations 6.31 and 6.34 imply

$$\left\|\prod_{i=j+1}^{\nu_{m+1}-1} (\mathbf{I} - \mathbf{a}_i\mathbf{h}_i')\right\| \le \left\|\prod_{i=\nu_{k+1}}^{\nu_{m+1}-1} (\mathbf{I} - \mathbf{a}_i\mathbf{h}_i')\right\| \left\|\prod_{i=j+1}^{\nu_{k+1}-1} (\mathbf{I} - \mathbf{a}_i\mathbf{h}_i')\right\| \le \prod_{i=k+1}^{m} \|\mathbf{Q}_i\| M^q,$$

where void products are to be read as unity. This, plus Equation 6.55, gives

$$\left\| \sum_{k=1}^{m} \sum_{j \in J_k} \prod_{i=j+1}^{\nu_{m+1}-1} (\mathbf{I} - \mathbf{a}_i \mathbf{h}_i')\mathbf{a}_j \mathbf{h}_j'(\mathbf{s}_j - \mathbf{s}) \right\|$$
$$\leq M^q \sum_{k=1}^{m} \prod_{i=k+1}^{m} \|\mathbf{Q}_i\| \sum_{j \in J_k} \|\mathbf{a}_j\| \, \|\mathbf{h}_j\| \, \|\mathbf{s}_j - \mathbf{s}\|$$
$$\leq M^q \sum_{k=1}^{m} \prod_{i=k+1}^{m} \|\mathbf{Q}_i\| \, \Delta_k d_k, \tag{6.59}$$

where, according to Equation 6.50,

$$d_k = \max_{j \in J_k} \|\mathbf{s}_j - \mathbf{s}\| \overset{\text{a.s.}}{\to} 0 \tag{6.60}$$

as $k \to \infty$. It is now a simple matter to show that the last sum in Equation 6.59 tends a.s. to zero as $m \to \infty$, which will complete the proof. Using Equation 6.57, we have

$$\sum_{k=1}^{N-1} \prod_{i=N+1}^{m} \|\mathbf{Q}_i\| \prod_{i=k+1}^{N} \|\mathbf{Q}_i\| \, \Delta_k d_k + \sum_{k=N}^{m} \prod_{i=k+1}^{m} \|\mathbf{Q}_i\| \, \Delta_k d_k$$
$$\leq C' \prod_{i=N+1}^{m} (1 - c\Delta_i) + \frac{1}{c} \sum_{k=N}^{m} b_{mk} d_k \tag{6.61}$$

after setting

$$c\Delta_k \prod_{i=k+1}^{m} (1 - c\Delta_i) = b_{mk}.$$

By Equation 6.58, $b_{mk} \overset{\text{a.s.}}{\to} 0$ as $m \to \infty$ for every fixed k; in particular, the first term in Equation 6.61 is a.s. $o(1)$ as $m \to \infty$. According to Lemma 1, we have

$$\sum_{k=N}^{m} b_{mk} = 1 - \prod_{k=N}^{m} (1 - c\Delta_k) \overset{\text{a.s.}}{\to} 1$$

as $m \to \infty$ for the same reason. It therefore follows from the Toeplitz Lemma (Knopp, 1947, p. 75) and Equation 6.60 that the second term in Equation 6.61 tends a.s. to zero as $m \to \infty$. Q.E.D.

6.4 Truncated Vector Iterations

The asymptotic behavior of truncated vector estimator sequences poses difficult analytical problems, and we have been only partially successful in dealing with them. However, because of their great practical importance, we feel that a discussion of truncated procedures is in order even though our results are not complete.

Suppose that prior information is available which allows us to state with certainty that the unknown $\boldsymbol{\theta}$ is an interior point of a closed (not necessarily bounded) convex subset, $\mathscr{P}$, of p-dimensional Euclidean space. (In the absence of prior knowledge, $\mathscr{P} = R^p$.) For any vector $\mathbf{x}$ in R^p, there is always a unique point $[\mathbf{x}]_{\mathscr{P}}$ in $\mathscr{P}$ which is closest to $\mathbf{x}$ in the sense of minimum Euclidean distance. In particular, if $\mathbf{x} \in \mathscr{P}$, then $[\mathbf{x}]_{\mathscr{P}} = \mathbf{x}$. If $\mathbf{x} \notin \mathscr{P}$, then $[\mathbf{x}]_{\mathscr{P}}$ is a boundary point of $\mathscr{P}$.

The natural generalization of the truncated scalar iteration of Chapters 2 through 5 is

$$\mathbf{t}_{n+1} = [\mathbf{t}_n + \mathbf{a}_n[Y_n - F_n(\mathbf{t}_n)]]_{\mathscr{P}},$$

where $\mathbf{t}_1$ is arbitrary in $\mathscr{P}$ (which reduces to Equation 2.2 if $p = 1$ and $\mathscr{P}$ is the interval $[\xi_1, \xi_2]$). In higher dimensions, the computational and analytical problem associated with actual evaluation of $[\mathbf{x}]_{\mathscr{P}}$ can be considerable. One must carry out the minimization of a quadratic form over a convex set and, if $\mathscr{P}$ is chosen inconveniently, this gets messy. Furthermore, it must be repeated for every new estimate. For this reason, one should, whenever possible, take $\mathscr{P}$ to be either a sphere or a rectangular parallelepiped whose sides are parallel to the coordinate planes. In these cases, the evaluation of $[\mathbf{x}]_{\mathscr{P}}$ is easy. If $\mathscr{P}$ is the sphere of radius r centered on a point $\boldsymbol{\theta}_0$, then

$$[\mathbf{x}]_{\mathscr{P}} = \min(r, \|\mathbf{x} - \boldsymbol{\theta}_0\|) \frac{\mathbf{x} - \boldsymbol{\theta}_0}{\|\mathbf{x} - \boldsymbol{\theta}_0\|} + \boldsymbol{\theta}_0.$$

If $\mathscr{P}$ is the rectangular parallelepiped $\{\mathbf{x}: \alpha_i \le x_i \le \beta_i, i = 1, 2, \cdots, p\}$, where x_i is the ith component of $\mathbf{x}$, then the ith component of the vector $[\mathbf{x}]_{\mathscr{P}}$ is simply $[x_i]_{\alpha_i}^{\beta_i}$, which was introduced in Equation 2.1.

We conjecture that the following proposition is a valid extension of Theorems 6.2 and 6.3, but the methods used to establish those results do not seem to work for the present situation.

6.5 Conjectured Theorem and Assumptions (E1 Through E6′)

CONJECTURED THEOREM

Let $\{Y_n: n = 1, 2, \cdots\}$ be a real-valued stochastic process of the form $Y_n = F_n(\boldsymbol{\theta}) + W_n$, where $F_n(\cdot)$ is known up to the p-dimensional parameter $\boldsymbol{\theta}$, and $W_1, W_2, \cdots$ have uniformly bounded variances. Let $\mathscr{P}$ be a closed convex subset of Euclidean p-space known to contain $\boldsymbol{\theta}$ as an interior point. For each n, let $\mathbf{a}_n(\cdot)$ be a Borel measurable mapping of $\mathscr{P}^{(n)} = \times_1^n \mathscr{P}$ into R^p, and let

$$\mathbf{t}_{n+1} = [\mathbf{t}_n + \mathbf{a}_n(\mathbf{t}_1, \cdots, \mathbf{t}_n)[Y_n - F_n(\mathbf{t}_n)]]_{\mathscr{P}}$$
$$(n = 1, 2, \cdots; \mathbf{t}_1 \text{ arbitrary in } \mathscr{P}).$$

Denote the gradient of F_n by $\dot{\mathbf{F}}_n$, and suppose the following conditions are met:

E1. $\lim_n \sup_{\mathbf{x}\in\mathscr{P}^{(n)},\mathbf{y}\in\mathscr{P}} \|\mathbf{a}_n(\mathbf{x})\| \, \|\dot{\mathbf{F}}_n(\mathbf{y})\| = 0.$

E2. $\sum_n \inf_{\mathbf{x}\in\mathscr{P}^{(n)},\mathbf{y}\in\mathscr{P}} \|\mathbf{a}_n(\mathbf{x})\| \, \|\dot{\mathbf{F}}_n(\mathbf{y})\| = \infty.$

E3. There exists a sequence of integers $1 = \nu_1 < \nu_2 < \nu_3 \cdots$ such that, with $p_k = \nu_{k+1} - \nu_k$,

$$p \le p_k \le q < \infty \qquad (k = 1, 2, \cdots),$$

and

$$\liminf_k \frac{1}{p_k} \inf_{(\mathbf{y}_{\nu_k},\cdots,\mathbf{y}_{\nu_{k+1}-1})\in\mathscr{P}^{(p_k)}} \lambda_{\min}\left(\sum_{j\in J_k} \frac{\dot{\mathbf{F}}_j(\mathbf{y}_j)\dot{\mathbf{F}}_j'(\mathbf{y}_j)}{\|\dot{\mathbf{F}}_j(\mathbf{y}_j)\|^2}\right) = \tau^2 > 0,$$

where

$$J_k = \{\nu_k, \nu_k + 1, \cdots, \nu_{k+1} - 1\}.$$

E4. $$\limsup_k \sup_{(\mathbf{x}_1,\cdots,\mathbf{x}_{\nu_{k+1}-1})\in\mathscr{P}^{(\nu_{k+1}-1)},\mathbf{y}\in\mathscr{P}} \frac{\max_{j\in J_k} \|\mathbf{a}_j(\mathbf{x}_1,\cdots,\mathbf{x}_j)\| \, \|\dot{\mathbf{F}}_j(\mathbf{y}_j)\|}{\min_{j\in J_k} \|\mathbf{a}_j(\mathbf{x}_1,\cdots,\mathbf{x}_j)\| \, \|\dot{\mathbf{F}}_j(\mathbf{y}_j)\|} = \rho < \infty.$$

E5. $$\liminf_n \inf_{\mathbf{x}\in\mathscr{P}^{(n)},\mathbf{y}\in\mathscr{P}} \frac{\mathbf{a}_n'(\mathbf{x}_1,\cdots,\mathbf{x}_n)\dot{\mathbf{F}}_n(\mathbf{y})}{\|\mathbf{a}_n(\mathbf{x}_1,\cdots,\mathbf{x}_n)\| \, \|\dot{\mathbf{F}}_n(\mathbf{y})\|} > \alpha = \sqrt{\frac{1-\tau^2}{1-\tau^2+(\tau/\rho)^2}}$$

Then $\mathbf{t}_n \to \boldsymbol{\theta}$ as $n \to \infty$ both in mean square and with probability one, if either

E6. $\sum_n \sup_{\mathbf{x}\in\mathscr{P}^{(n)}} \|\mathbf{a}_n(\mathbf{x})\| < \infty,$

or

E6′. $\{W_n\}$ is a zero-mean independent process and

$$\sum_n \sup_{\mathbf{x}\in\mathscr{P}^{(n)}} \|\mathbf{a}_n(\mathbf{x})\|^2 < \infty.$$

In practice, Assumptions E1 through E6′ represent a substantial relaxation of Assumptions C1 through C6′, particularly when $\mathscr{P}$ is a small compact set.

6.6 Batch Processing

One of the reasons we feel so strongly about the veracity of the conjecture is that the part of the proposition concerning mean-square convergence is true if the estimation scheme is only slightly modified; specifically, if the data are collected and processed in batches. Actually, batch processing is sometimes the most natural way to handle an incoming flow of data. For example, if information is being collected simultaneously from many sources (say, p_k sources reporting at time k), then it is reasonable to process all the currently available data before updating the most recent estimate. A less convincing, but still plausible, case can be made for batch-processing data that arrive in a steady stream. In either case, the raw data consist of scalar observations $Y_n = F_n(\theta) + W_n$ $(n = 1, 2, \cdots)$. These are grouped together so that the data processed at the kth instant is the vector of observations

$$\mathbf{Y}_k = \begin{bmatrix} Y_{\nu_k} \\ Y_{\nu_k+1} \\ \vdots \\ Y_{\nu_{k+1}-1} \end{bmatrix} = \mathbf{F}_k(\boldsymbol{\theta}) + \mathbf{W}_k \qquad (k = 1, 2, \cdots),$$

where the regression vector $\mathbf{F}_k(\boldsymbol{\theta})$ and the residual vector $\mathbf{W}_k$ each have p_k components and are defined in the obvious way.

The recursion considered is of the form

$$\mathbf{s}_{k+1} = \mathbf{s}_k + \mathbf{A}_k[\mathbf{Y}_k - \mathbf{F}_k(\mathbf{s}_k)],$$

where $\mathbf{A}_k$ is a suitably chosen $p \times p_k$ matrix. We now define

$$\mathbf{t}_n = \mathbf{s}_k \qquad \text{for all } n \in J_k.$$

This vector estimate "keeps pace" with the incoming data, but instead of changing after each observation, it changes only after $p_1, p_2, \cdots$ observations.

The mean-square consistency of truncated recursive estimators subjected to batch processing is the substance of the next theorem.

THEOREM 6.4

Let $\{Y_n\}$, $\{\mathbf{a}_n(\cdot)\}$, and $\mathscr{P}$ be as defined in the previous statement of the Conjectured Theorem, and suppose that the Assumptions E1 through E5 and either E6 or E6′ hold. Let $\mathbf{s}_1$ be arbitrary in $\mathscr{P}$, and let

$$\mathbf{s}_{k+1} = [\mathbf{s}_k + \mathbf{A}_k[\mathbf{Y}_k - \mathbf{F}_k(\mathbf{s}_k)]]_{\mathscr{P}} \qquad (k = 1, 2, \cdots),$$

where

$$\mathbf{Y}_k = \begin{bmatrix} Y_{\nu_k} \\ \vdots \\ Y_{\nu_{k+1}-1} \end{bmatrix}, \qquad \mathbf{F}_k(\mathbf{x}) = \begin{bmatrix} F_{\nu_k}(\mathbf{x}) \\ \vdots \\ F_{\nu_{k+1}-1}(\mathbf{x}) \end{bmatrix}.$$

Here $\mathbf{A}_k$ is the matrix whose columns are

$$\mathbf{a}_{\nu_k}, \cdots, \mathbf{a}_{\nu_{k+1}-1},$$

where $\mathbf{a}_j = \mathbf{a}_j(\mathbf{t}_1, \cdots, \mathbf{t}_j)$ and

$$\mathbf{t}_j = \mathbf{s}_k \qquad \text{for all} \quad j \in J_k.$$

Then $\mathscr{E}\|\mathbf{s}_k - \boldsymbol{\theta}\|^2 \to 0$ as $k \to \infty$.

Proof. We define the p-vectors

$$\mathbf{S}_k = \mathbf{s}_k + \mathbf{A}_k[\mathbf{F}_k(\boldsymbol{\theta}) - \mathbf{F}_k(\mathbf{s}_k)], \qquad \mathbf{Z}_k = \mathbf{A}_k[\mathbf{Y}_k - \mathbf{F}_k(\boldsymbol{\theta})] = \mathbf{A}_k\mathbf{W}_k, \tag{6.62}$$

where $\mathbf{W}_k$ has the p_k components $W_{\nu_k}, W_{\nu_k+1}, \cdots, W_{\nu_{k+1}-1}$. Then we have

$$\mathbf{s}_{k+1} = [\mathbf{S}_k + \mathbf{Z}_k]_{\mathscr{P}}.$$

If $\mathbf{S}_k + \mathbf{Z}_k \in \mathscr{P}$, then $\|\mathbf{s}_{k+1} - \boldsymbol{\theta}\| = \|\mathbf{S}_k + \mathbf{Z}_k - \boldsymbol{\theta}\|$. If $\mathbf{S}_k + \mathbf{Z}_k$ does not belong to $\mathscr{P}$, the hyperplane that passes through the boundary point $\mathbf{s}_{k+1}$ and is perpendicular to the line joining $\mathbf{s}_{k+1}$ and $\mathbf{S}_k + \mathbf{Z}_k$ separates the latter from all points in $\mathscr{P}$. (This is the classical way of constructing a hyperplane that separates a convex body from any point outside.) Thus, all points in $\mathscr{P}$, $\boldsymbol{\theta}$ in particular, are closer to $\mathbf{s}_{k+1}$ than to $\mathbf{S}_k + \mathbf{Z}_k$. Consequently, we see that

$$\|\mathbf{s}_{k+1} - \boldsymbol{\theta}\| \le \|\mathbf{S}_k + \mathbf{Z}_k - \boldsymbol{\theta}\|$$

is true in either case.

By the mean-value theorem, we have

$$\mathbf{F}_k(\mathbf{s}_k) = \mathbf{F}_k(\boldsymbol{\theta}) + \mathbf{H}_k'(\mathbf{s}_k - \boldsymbol{\theta}_k),$$

where $\mathbf{H}_k$ is the matrix whose columns are, respectively, the gradient vectors of $F_{\nu_k}, F_{\nu_k+1}, \cdots, F_{\nu_{k+1}-1}$ evaluated at various points in $\mathscr{P}$. Thus, we have

$$\mathbf{S}_k - \boldsymbol{\theta} = (\mathbf{I} - \mathbf{A}_k\mathbf{H}_k')(\mathbf{s}_k - \boldsymbol{\theta}),$$

and

$$\|\mathbf{s}_{k+1} - \boldsymbol{\theta}\| \le \|(\mathbf{I} - \mathbf{A}_k\mathbf{H}_k')(\mathbf{s}_k - \boldsymbol{\theta}) + \mathbf{Z}_k\|.$$

From this and Equation 6.62, we deduce the inequality

$$\|\mathbf{s}_{k+1} - \boldsymbol{\theta}\|^2 \le \|\mathbf{I} - \mathbf{A}_k\mathbf{H}_k'\|^2\|\mathbf{s}_k - \boldsymbol{\theta}\|^2 + 2\mathbf{W}_k'\mathbf{A}_k'(\mathbf{I} - \mathbf{A}_k\mathbf{H}_k')(\mathbf{s}_k - \boldsymbol{\theta}) + \|\mathbf{A}_k\mathbf{W}_k\|^2. \tag{6.63}$$

If we can show that

$$\|\mathbf{I} - \mathbf{A}_k\mathbf{H}_k'\|^2 \le (1 - c\Delta_k)^2 \tag{6.64}$$

holds for some (deterministic) $c > 0$ and number sequence $\{\Delta_k\}$ such that

$$\Delta_k \to 0, \qquad \Sigma\Delta_k = \infty, \tag{6.65}$$

then we will be finished.

Indeed, if we set

$$e_k^2 = \mathscr{E}\|\mathbf{s}_k - \boldsymbol{\theta}\|^2,$$

it follows from Equation 6.63, after first majorizing the middle term with norms, that

$$e_{k+1}^2 \leq (1 - c\Delta_k)^2 e_k^2 + M_1 \mathscr{E}^{1/2}(\|\mathbf{A}_k\| \, \|\mathbf{W}_k\|)^2 e_k + \mathscr{E}(\|\mathbf{A}_k\| \, \|\mathbf{W}_k\|)^2. \tag{6.66}$$

Since

$$\begin{aligned} \|\mathbf{A}_k\|^2 &\leq \operatorname{tr}(\mathbf{A}_k'\mathbf{A}_k) = \sum_{j \in J_k} \|\mathbf{a}_j\|^2 \\ &\leq \sum_{j \in J_k} \sup_{\mathbf{x} \in \mathscr{S}(j)} \|\mathbf{a}_j(\mathbf{x})\|^2, \end{aligned} \tag{6.67}$$

it follows that

$$\|\mathbf{A}_k\| \leq q \max_{j \in J_k} \sup_{\mathbf{x} \in \mathscr{S}(j)} \|\mathbf{a}_j(\mathbf{x})\| = \alpha_k. \tag{6.68}$$

Under Assumption E6,

$$\Sigma\alpha_k < \infty. \tag{6.69}$$

By hypothesis, all residual second moments are finite, so from Equation 6.66 it follows that

$$e_{k+1}^2 \leq (1 - c\Delta_k)^2 e_k^2 + M_2\alpha_k e_k + M_3\alpha_k^2 \leq (1 - c\Delta_k)^2 e_k^2 + M_4\alpha_k(1 + e_k).$$

According to Lemma 3, Equations 6.65 and 6.69 imply $\sup_k e_k^2 < \infty$; therefore,

$$e_{k+1}^2 \leq \left(1 - \frac{3}{2} c\Delta_k\right) e_k^2 + M_5\alpha_k.$$

The argument that followed Equation 6.27 shows that $e_k^2 \to 0$ as $k \to \infty$. On the other hand, when Assumption E6′ is true, the cross term in Equation 6.63 has zero expectation, because $\mathbf{W}_k$ is independent of $\mathbf{A}_k$, $\mathbf{H}_k$, and $\mathbf{s}_k$. We now find that

$$e_{k+1}^2 \leq \left(1 - \frac{3}{2} c\Delta_k\right) e_k^2 + M_6\alpha_k^2.$$

The sequence of Equation 6.68 is square summable under Assumption E6′, so again e_k^2 tends to zero.

It remains, therefore, to establish Equations 6.64 through 6.65. Let $\mathbf{h}_{\nu_k}, \cdots, \mathbf{h}_{\nu_{k+1}-1}$ be the columns of $\mathbf{H}_k$, so that $\mathbf{h}_j = \dot{\mathbf{F}}(\mathbf{u}_j)$ for some $\mathbf{u}_j \in \mathscr{P}$. Then we have

$$\mathbf{A}_k\mathbf{H}_k' = \sum_{j \in J_k} \mathbf{a}_j\mathbf{h}_j'$$

and

$$\begin{aligned}
\|\mathbf{I} - \mathbf{A}_k\mathbf{H}_k'\|^2 &= \|(\mathbf{I} - \mathbf{A}_k\mathbf{H}_k')'(\mathbf{I} - \mathbf{A}_k\mathbf{H}_k)\| \\
&\le \left\|\mathbf{I} - \sum_{j \in J_k} (\mathbf{a}_j\mathbf{h}_j' + \mathbf{h}_j\mathbf{a}_j')\right\| + \|\mathbf{A}_k\mathbf{H}_k'\|^2 \\
&\le \left\|\mathbf{I} - \sum_{j \in J_k} (\mathbf{a}_j\mathbf{h}_j' + \mathbf{h}_j\mathbf{a}_j')\right\| + \sum_{j \in J_k} \|\mathbf{a}_j\|^2\|\mathbf{h}_j\|^2. \qquad (6.70)
\end{aligned}$$

We set

$$\Delta_k\mathbf{T}_k = \sum_{j \in J_k} (\mathbf{a}_j\mathbf{h}_j' + \mathbf{h}_j'\mathbf{a}_j),$$

where

$$\Delta_k = \left(\sum_{j \in J_k} \inf_{\mathbf{x} \in \mathscr{P}^{(j)}, \mathbf{y} \in \mathscr{P}} \|\mathbf{a}_j(\mathbf{x})\|^2\|\dot{\mathbf{F}}(\mathbf{y})\|^2\right)^{1/2}. \qquad (6.71)$$

The latter satisfies Equation 6.65, according to Assumptions E1 and E2. Furthermore, Equation 6.70 and Assumption E4 give

$$\begin{aligned}
\|\mathbf{I} - \mathbf{A}_k\mathbf{H}_k'\|^2 &\le \|\mathbf{I} - \Delta_k(\mathbf{T}_k + \mathbf{T}_k')\| + M\Delta_k^2 \\
&= 1 - \Delta_k\lambda_{\min}(\mathbf{T}_k + \mathbf{T}_k') + M\Delta_k^2.
\end{aligned}$$

The same argument used in the proof of Theorem 6.1 applies here, under Assumptions E3 through E5, to prove that

$$\lambda_{\min}(\mathbf{T}_k + \mathbf{T}_k') > 3c > 0$$

for all large enough k, and hence Equation 6.64. Q.E.D.

The following is a rather trivial corollary to Theorem 6.4, but it will prove useful in the discussion of certain applications.

Corollary. If the sequence of gain vectors $\{\mathbf{a}_n(\cdot)\}$ satisfies E1, E2, E4, E5, and either E6 or E6′, then so does

$$\mathbf{a}_n^*(\cdot) = \varphi_n\mathbf{a}_n(\cdot)$$

for any sequence $\{\varphi_n\}$ of scalars bounded from above and below away from zero.

7. Complements and Details

In this chapter we will examine various rationales for choosing gain sequences for vector-parameter recursive estimation schemes. Motivated by consideration of the linear case, two types of gains will be discussed in detail. The first category of gains possesses an optimal property when applied to linear regression. The other has the virtue of extreme computational simplicity. The results of Theorem 6.4 are specialized and applied directly to these particular gains in Theorem 7.1. We begin our discussion with a look at linear regression from the recursive point of view.

7.1 Optimum Gains for Recursive Linear Regression

Suppose one observes random variables,

$$Y_n = \mathbf{h}_n'\boldsymbol{\theta} + W_n, \tag{7.1}$$

where $\{\mathbf{h}_n\}$ is a known sequence of p-dimensional vectors, $\boldsymbol{\theta}$ is not known, and the W_n's are assumed to be independent random variables with common unknown variances σ^2. Suppose further that someone presents us with an estimator $\mathbf{t}_n$ that is based upon (is a measurable function of) the first $n - 1$ observations. We construct an estimator $\mathbf{t}_{n+1}$ that incorporates the nth observation Y_n in the following way:

$$\mathbf{t}_{n+1} = \mathbf{t}_n + \mathbf{a}_n[Y_n - \mathbf{h}_n'\mathbf{t}_n]. \tag{7.2}$$

How should the (deterministic) gain vector $\mathbf{a}_n$ be chosen if we wish to minimize the mean-square distance from $\mathbf{t}_{n+1}$ to $\boldsymbol{\theta}$?

This question can be answered in a straightforward manner. Let

$$\mathbf{B}_n = \frac{1}{\sigma^2}\mathscr{E}(\mathbf{t}_{n+1} - \boldsymbol{\theta})(\mathbf{t}_{n+1} - \boldsymbol{\theta})' \qquad (n = 1, 2, \cdots).$$

Since

$$\mathscr{E}\|\mathbf{t}_{n+1} - \boldsymbol{\theta}\|^2 = \operatorname{tr}\mathscr{E}(\mathbf{t}_{n+1} - \boldsymbol{\theta})(\mathbf{t}_{n+1} - \boldsymbol{\theta})', \tag{7.3}$$

it is clear that $\mathbf{a}_n$ should be chosen to minimize the trace of $\mathbf{B}_n$. Substituting Equation 7.1 into Equation 7.2, exploiting the independence of the W's, and completing the square, we find that

$$\mathbf{B}_n = \mathbf{B}_{n-1} - \frac{(\mathbf{B}_{n-1}\mathbf{h}_n)(\mathbf{B}_{n-1}\mathbf{h}_n)'}{1 + \mathbf{h}_n'\mathbf{B}_{n-1}\mathbf{h}_n}$$
$$+ (1 + \mathbf{h}_n'\mathbf{B}_{n-1}\mathbf{h}_n)\left(\mathbf{a}_n - \frac{\mathbf{B}_{n-1}\mathbf{h}_n}{1 + \mathbf{h}_n'\mathbf{B}_{n-1}\mathbf{h}_n}\right)\left(\mathbf{a}_n - \frac{\mathbf{B}_{n-1}\mathbf{h}_n}{1 + \mathbf{h}_n'\mathbf{B}_{n-1}\mathbf{h}_n}\right)'. \tag{7.4}$$

Thus,

$$\operatorname{tr}\mathbf{B}_n = \operatorname{tr}\mathbf{B}_{n-1} - \frac{\mathbf{h}_n'\mathbf{B}_{n-1}^2\mathbf{h}_n}{1 + \mathbf{h}_n'\mathbf{B}_{n-1}\mathbf{h}_n}$$
$$+ (1 + \mathbf{h}_n'\mathbf{B}_{n-1}\mathbf{h}_n)\left\|\mathbf{a}_n - \frac{\mathbf{B}_{n-1}\mathbf{h}_n}{1 + \mathbf{h}_n'\mathbf{B}_{n-1}\mathbf{h}_n}\right\|^2. \tag{7.5}$$

Thus, if the estimator $\mathbf{t}_n$ is given (with second-order moment matrix $\mathbf{B}_{n-1}$), the appropriate value of $\mathbf{a}_n$ (which minimizes tr $\mathbf{B}_n$) is given by

$$\mathbf{a}_n = \frac{\mathbf{B}_{n-1}\mathbf{h}_n}{1 + \mathbf{h}_n'\mathbf{B}_{n-1}\mathbf{h}_n}. \tag{7.6}$$

When $\mathbf{a}_n$ is so chosen,

$$\mathbf{B}_n = \mathbf{B}_{n-1} - \frac{(\mathbf{B}_{n-1}\mathbf{h}_n)(\mathbf{B}_{n-1}\mathbf{h}_n)'}{1 + \mathbf{h}_n'\mathbf{B}_{n-1}\mathbf{h}_n}, \tag{7.7}$$

and

$$\mathbf{B}_n\mathbf{h}_n = \mathbf{B}_{n-1}\mathbf{h}_n\left(1 - \frac{\mathbf{h}_n'\mathbf{B}_{n-1}\mathbf{h}_n}{1 + \mathbf{h}_n'\mathbf{B}_{n-1}\mathbf{h}_n}\right)$$
$$= \frac{\mathbf{B}_{n-1}\mathbf{h}_n}{1 + \mathbf{h}_n'\mathbf{B}_{n-1}\mathbf{h}_n} = \mathbf{a}_n.$$

Thus, the same end is achieved by choosing

$$\mathbf{a}_n = \mathbf{B}_n\mathbf{h}_n, \tag{7.8}$$

where $\mathbf{B}_n$ is defined in terms of $\mathbf{B}_{n-1}$ by Equation 7.7. This result leads us to give serious consideration to gain sequences defined iteratively by

Equations 7.7 and 7.8, at least insofar as recursive estimation procedures for linear regression are concerned. Let us, therefore, consider in detail the recursive estimation scheme

$$\begin{aligned} \mathbf{t}_{n+1} &= \mathbf{t}_n + \mathbf{a}_n[Y_n - \mathbf{h}_n'\mathbf{t}_n], \\ \mathbf{a}_n &= \mathbf{B}_n\mathbf{h}_n, \\ \mathbf{B}_n &= \mathbf{B}_{n-1} - \frac{(\mathbf{B}_{n-1}\mathbf{h}_n)(\mathbf{B}_{n-1}\mathbf{h}_n)'}{1 + \mathbf{h}_n'\mathbf{B}_{n-1}\mathbf{h}_n}. \end{aligned} \tag{7.9}$$

In order to "get the recursion started," initial conditions for $\mathbf{t}_n$ and $\mathbf{B}_{n-1}$ must be specified for some $n = n_0 + 1$. If this is done, it is easy to show that the recursions can be solved in closed form:

$$\mathbf{B}_{n_0+k} = \left(\mathbf{B}_{n_0}^{-1} + \sum_{j=1}^{k} \mathbf{h}_{n_0+j}\mathbf{h}_{n_0+j}'\right)^{-1} \quad (k = 0, 1, 2, \cdots), \tag{7.10}$$

and

$$\mathbf{t}_{n+1} = \mathbf{B}_n\left(\mathbf{B}_{n_0}^{-1}\mathbf{t}_{n_0+1} + \sum_{j=n_0+1}^{n} \mathbf{h}_j Y_j\right), \quad (n \geq n_0), \tag{7.11}$$

provided that $\mathbf{B}_{n_0}$ is positive definite. To prove Equation 7.10, we proceed by induction. It is true for $k = 0$. Suppose it holds for k. Then

$$\left(\mathbf{B}_{n_0}^{-1} + \sum_{j=1}^{k+1} \mathbf{h}_{n_0+j}\mathbf{h}_{n_0+j}'\right)\mathbf{B}_{n_0+k+1} = (\mathbf{B}_{n_0+k}^{-1} + \mathbf{h}_{n_0+k+1}\mathbf{h}_{n_0+k+1}')\mathbf{B}_{n_0+k+1}. \tag{7.12}$$

Now we use Equation 7.9, write $\mathbf{B}_{n_0+k+1}$ in terms of $\mathbf{B}_{n_0+k}$ and $\mathbf{h}_{n_0+k+1}$, and carry out the multiplication on the right-hand side of Equation 7.12. We thereby obtain the identity matrix, and this establishes Equation 7.10. Equation 7.11 is verified by substitution.

Let us examine the consequences of some special initial conditions. If we take $n_0 = 0$, $\mathbf{t}_1 = \mathbf{0}$, and $\mathbf{B}_0 = \mathbf{R}_0$, then

$$\mathbf{B}_n = \left(\mathbf{R}_0^{-1} + \sum_{j=1}^{n} \mathbf{h}_j\mathbf{h}_j'\right)^{-1} \tag{7.13}$$

and

$$\mathbf{t}_{n+1} = \left(\mathbf{R}_0^{-1} + \sum_{j=1}^{n} \mathbf{h}_j\mathbf{h}_j'\right)^{-1}\left(\sum_{j=1}^{n} Y_j\mathbf{h}_j\right) \tag{7.14}$$

This is exactly the expression for the conditional expectation of $\boldsymbol{\theta}$, given $Y_1, \cdots, Y_n$, in the case where the residuals have a spherically symmetric joint normal distribution with variance σ^2, and $\boldsymbol{\theta}$ has a prior normal distribution with zero mean and covariance $\sigma^2\mathbf{R}_0$.

Suppose, on the other hand, that we wait for p observations to accumulate before attempting to estimate the p-dimensional parameter $\boldsymbol{\theta}$. If we assume that $\mathbf{h}_1, \mathbf{h}_2, \cdots, \mathbf{h}_p$ are linearly independent and take, as our "first" estimate,

$$\mathbf{t}_{p+1} = \left(\sum_{j=1}^{p} \mathbf{h}_j \mathbf{h}_j'\right)^{-1} \left(\sum_{j=1}^{p} \mathbf{h}_j Y_j\right),$$

then, since $Y_j = \mathbf{h}_j'\boldsymbol{\theta} + W_j$, we see that

$$(\mathbf{t}_{p+1} - \boldsymbol{\theta}) = \left(\sum_{j=1}^{p} \mathbf{h}_j \mathbf{h}_j'\right)^{-1} \left(\sum_{j=1}^{p} \mathbf{h}_j W_j\right).$$

We have, therefore,

$$\mathscr{E}(\mathbf{t}_{p+1} - \boldsymbol{\theta})(\mathbf{t}_{p+1} - \boldsymbol{\theta})' = \sigma^2 \left(\sum_{j=1}^{p} \mathbf{h}_j \mathbf{h}_j'\right)^{-1}.$$

Thus, if we take

$$n_0 = p, \qquad \text{and} \qquad \mathbf{B}_p = \left(\sum_{j=1}^{p} \mathbf{h}_j \mathbf{h}_j'\right)^{-1},$$

we deduce from Equation 7.10 that

$$B_n = \left(\sum_{j=1}^{n} \mathbf{h}_j \mathbf{h}_j'\right)^{-1} \qquad \text{for} \quad n \geq p.$$

Furthermore, by Equation 7.11,

$$\mathbf{t}_{n+1} = \mathbf{B}_n \left(\sum_{j=1}^{p} \mathbf{h}_j Y_j + \sum_{j=p+1}^{n} \mathbf{h}_j Y_j\right)$$

$$= \left(\sum_{j=1}^{n} \mathbf{h}_j \mathbf{h}_j'\right)^{-1} \left(\sum_{j=1}^{n} \mathbf{h}_j Y_j\right), \tag{7.15}$$

which is precisely the least-squares estimator for $\boldsymbol{\theta}$ based upon $Y_1, Y_2, \cdots, Y_n$.

In the more conventional matrix notation, the Bayesian and least-squares estimators, Equations 7.14 and 7.15, can be written as

$$\mathbf{t}_{n+1} = (\mathbf{H}_n \mathbf{H}_n' + \mathbf{R}_0^{-1})^{-1} \mathbf{H}_n \mathbf{Y}_n, \qquad \text{and} \qquad \mathbf{t}_{n+1} = (\mathbf{H}_n \mathbf{H}_n')^{-1} \mathbf{H}_n \mathbf{Y}_n,$$

respectively, where $\mathbf{H}_n'$ is the $n \times p$ matrix whose rows are $\mathbf{h}_j'$ ($j = 1, 2, \cdots, n$), and $\mathbf{Y}_n$ is the n vector whose jth component is the scalar observation Y_j. Thus, depending upon the initial conditions, the recursion of Equation 7.9 can yield the Bayesian estimator of $\boldsymbol{\theta}$ (conditional expectation) in a Gaussian formulation, or the least-squares estimator for $\boldsymbol{\theta}$ (no assumptions concerning distribution theory of residuals being necessary).

From the large-sample point of view, the initial conditions are of no consequence. In fact, by Equation 7.11,

$$\mathscr{E}(\mathbf{t}_{n+1} - \boldsymbol{\theta})(\mathbf{t}_{n+1} - \boldsymbol{\theta})' = \mathbf{B}_n[\mathbf{A}_{n_0} + \sigma^2\mathbf{B}_n^{-1}]\mathbf{B}_n,$$

where

$$\mathbf{A}_{n_0} = \mathscr{E}\left(\mathbf{B}_{n_0}^{-1}\mathbf{t}_{n_0} - \sum_{j=1}^{n_0} \mathbf{h}_j Y_j\right)\left(\mathbf{B}_{n_0}^{-1}\mathbf{t}_{n_0} - \sum_{j=1}^{n_0} \mathbf{h}_j Y_j\right)$$

Therefore,

$$\begin{aligned}\mathscr{E}\|\mathbf{t}_{n+1} - \boldsymbol{\theta}\|^2 &= \operatorname{tr} \mathbf{B}_n[\mathbf{A}_{n_0} + \sigma^2\mathbf{B}_n^{-1}]\mathbf{B}_n \\ &= \operatorname{tr} \mathbf{B}_n\mathbf{A}_{n_0}\mathbf{B}_n + \sigma^2 \operatorname{tr} \mathbf{B}_n\end{aligned}$$

approaches zero if and only if $\operatorname{tr} \mathbf{B}_n \to 0$. Since

$$\lambda_{\max}(\mathbf{B}_n) \leq \operatorname{tr} \mathbf{B}_n \leq p\lambda_{\max}(\mathbf{B}_n),$$

this reduces the question of $\mathbf{t}_n$'s mean-square consistency to the study of $\mathbf{B}_n$'s largest eigenvalue. (We could resort to Theorem 6.2, but we will see that the special features of this linear problem make the hypotheses of Theorem 6.2 unnecessarily strong.) By Equation 7.10,

$$\begin{aligned}\mathbf{B}_n &= \left(\mathbf{B}_{n_0}^{-1} - \sum_{j=1}^{n_0} \mathbf{h}_j\mathbf{h}_j' + \sum_{j=1}^{n} \mathbf{h}_j\mathbf{h}_j'\right)^{-1} \\ &= \left(\mathbf{C}_{n_0} + \sum_{j=1}^{n} \mathbf{h}_j\mathbf{h}_j'\right)^{-1}.\end{aligned}$$

Since $\lambda_{\max}(\mathbf{B}_n) = 1/\lambda_{\min}(\mathbf{B}_n^{-1})$, and since $\lambda_{\min}(\mathbf{B}_n^{-1}) \to \infty$ if and only if

$$\mathbf{x}'\mathbf{B}_n^{-1}\mathbf{x} = \mathbf{x}'\mathbf{C}_{n_0}\mathbf{x} + \mathbf{x}'\left(\sum_{j=1}^{n} \mathbf{h}_j\mathbf{h}_j'\right)\mathbf{x} \to \infty$$

for every unit vector $\mathbf{x}$, we must find conditions which ensure that

$$\lim_n \lambda_{\min}\left(\sum_{j=1}^{n} \mathbf{h}_j\mathbf{h}_j'\right) = \infty. \tag{7.16}$$

Equation 7.16 will hold if there is a sequence of integers $1 = \nu_1 < \nu_2 < \cdots$, with

$$p \leq p_k = \nu_{k+1} - \nu_k \leq q < \infty \quad \text{and} \quad J_k = \{\nu_k, \nu_k + 1, \cdots, \nu_{k+1} - 1\},$$

such that

$$\liminf_{k\to\infty} \lambda_{\min}\left(\sum_{j\in J_k} \mathbf{h}_j\mathbf{h}_j'/\|\mathbf{h}_j\|^2\right) > 0 \tag{7.17a}$$

and

$$\sum_{n=1}^{\infty} \min_{j \in J_n} \|\mathbf{h}_j\|^2 = \infty. \tag{7.17b}$$

For then

$$\begin{aligned}
\lambda_{\min}\left(\sum_{j=1}^{\nu_{k+1}-1} \mathbf{h}_j\mathbf{h}_j'\right) &= \lambda_{\min}\left(\sum_{n=1}^{k} \sum_{j \in J_n} \mathbf{h}_j\mathbf{h}_j'\right) \\
&\geq \sum_{n=1}^{k} \lambda_{\min}\left(\sum_{j \in J_n} \mathbf{h}_j\mathbf{h}_j'\right) \\
&\geq \sum_{n=1}^{k} \min_{j \in J_n} \|\mathbf{h}_j\|^2 \lambda_{\min}\left(\sum_{j \in J_n} \mathbf{h}_j\mathbf{h}_j'/\|\mathbf{h}_j\|^2\right) \\
&\geq \tau^2 \sum_{n=1}^{k} \min_{j \in J_n} \|\mathbf{h}_j\|^2 \to \infty.
\end{aligned}$$

Since

$$s_n = \lambda_{\min}\left(\sum_{j=1}^{n} \mathbf{h}_j\mathbf{h}_j'\right)$$

is a nondecreasing sequence, it converges to the same limit as does $s_{\nu_{k+1}-1}$. This establishes Equation 7.16.

As we mentioned earlier, the special nature of the gains being used in this linear-regression problem causes some of the conditions of Theorem 6.2 to be irrelevant. In fact, Assumptions C2, C3, and C4 are by themselves sufficient to guarantee Equation 7.17. Quite obviously, Assumption C3 is the same as Equation 7.17a, and Assumption C4 implies that

$$\begin{aligned}
\min_{j \in J_k} \|\mathbf{B}_j\mathbf{h}_j\| \, \|\mathbf{h}_j\| &\geq \frac{1}{\rho} \max_{j \in J_k} \|\mathbf{B}_j\mathbf{h}_j\| \, \|\mathbf{h}_j\| \\
&\geq \frac{1}{\rho q} \sum_{j \in J_k} \|\mathbf{B}_j\mathbf{h}_j\| \, \|\mathbf{h}_j\|;
\end{aligned}$$

therefore, we have

$$\sum_k \min_{j \in J_k} \|\mathbf{B}_j\mathbf{h}_j\| \, \|\mathbf{h}_j\| \geq \frac{1}{\rho q} \sum_k \sum_{j \in J_k} \|\mathbf{B}_j\mathbf{h}_j\| \, \|\mathbf{h}_j\| = \infty$$

by Assumption C2. Since

$$\|\mathbf{B}_j\| = \lambda_{\max}(\mathbf{B}_j) = 1/\lambda_{\min}(\mathbf{B}_j^{-1}),$$

and since

$$\lambda_{\min}(\mathbf{B}_j^{-1}) = \lambda_{\min}\left(\sum_{i=1}^{j} \mathbf{h}_i\mathbf{h}_i'\right)$$

is a positive nondecreasing sequence, we see that

$$\limsup_{j\to\infty} \|\mathbf{B}_j\| < M.$$

Thus,

$$\infty = \sum_k \min_{j\in J_k} \|\mathbf{B}_j\mathbf{h}_j\|\,\|\mathbf{h}_j\| \le M \sum_k \min_{j\in J_k} \|\mathbf{h}_j\|^2,$$

which implies Equation 7.17*b*.

In summary, Assumptions C2, C3, and C4 of Theorem 6.2 are sufficient for Equation 7.17, which in turn implies Equation 7.16. This latter condition, in turn, is necessary and sufficient for the recursion of Equation 7.9 to generate a mean-square consistent estimator for the linear-regression parameter $\boldsymbol{\theta}$, *regardless of the initial conditions imposed upon the* $\mathbf{t}_n$ *and* $\mathbf{B}_n$ *sequences.*

7.2 "Quick and Dirty" Recursive Linear Regression

In situations where data pour in at a very high rate and "real-time" estimates are acutely desired, we may be willing to trade off statistical efficiency for computational speed, so long as consistency is preserved. The gain sequence

$$\mathbf{a}_n = \frac{\mathbf{h}_n}{\sum_{j=1}^{n} \|\mathbf{h}_j\|^2}, \tag{7.18}$$

with the associated recursion

$$\mathbf{t}_{n+1} = \mathbf{t}_n + \left[\frac{\mathbf{h}_n}{\sum_{j=1}^{n} \|\mathbf{h}_j\|^2}\right] [Y_n - \mathbf{h}_n'\mathbf{t}_n] \qquad (\mathbf{t}_1 \text{ arbitrary})$$

furnishes a very handy estimation scheme. The gains used in Equation 7.9 allowed us to find a closed form expression for $\mathbf{t}_n$ and to study its asymptotic properties directly (without recourse to Theorem 6.2). This is not possible in the present case. However, if we assume that Equation 7.17 holds and, in addition, that

$$0 < \liminf_n \frac{\|\mathbf{h}_{n+1}\|}{\|\mathbf{h}_n\|} \le \limsup_n \frac{\|\mathbf{h}_{n+1}\|}{\|\mathbf{h}_n\|} < \infty \tag{7.19a}$$

and

$$\lim_n \frac{\|\mathbf{h}_n\|^2}{\sum_{j=1}^{n} \|\mathbf{h}_j\|^2} = 0, \tag{7.19b}$$

then Assumptions C1 through C5 and C6′ are satisfied. To see why this is so, we begin by pointing out that in the present case

$$\|\mathbf{a}_n\| \, \|\dot{\mathbf{F}}_n\| = \frac{\|\mathbf{h}_n\|^2}{\sum_{j=1}^{n} \|\mathbf{h}_j\|^2} \to 0$$

by Equation 7.19*b*, which guarantees Assumption C1. By Equation 7.17*b*, we have $\sum_n \|\mathbf{h}_j\|^2 = \infty$; therefore, by the Abel–Dini Theorem (Equation 2.27, Chapter 2),

$$\sum_n \|\mathbf{a}_n\| \, \|\dot{\mathbf{F}}_n\| = \sum_n \left(\frac{\|\mathbf{h}_n\|^2}{\sum_{j=1}^{n} \|\mathbf{h}_j\|^2} \right) = \infty, \tag{7.20}$$

which establishes Assumption C2. Assumption C3 is the same as Equation 7.17*a*, while Equation 7.19*a* implies that

$$K_2\|\mathbf{h}_n\| \le \|\mathbf{h}_{n+1}\| \le K_1\|\mathbf{h}_n\|;$$

therefore, if $i, j \in J_n$,

$$K_2{}^q\|\mathbf{h}_j\| \le \|\mathbf{h}_i\| \le K_1{}^q\|\mathbf{h}_j\|.$$

Consequently

$$\max_{j \in J_k} \frac{\|\mathbf{h}_j\|^2}{\sum_{i=1}^{j} \|\mathbf{h}_i\|^2} \le \frac{K_1{}^q\|\mathbf{h}_{\nu_k}\|^2}{\sum_{i=1}^{\nu_k} \|\mathbf{h}_i\|^2},$$

while

$$\min_{j \in J_k} \frac{\|\mathbf{h}_j\|^2}{\sum_{i=1}^{j} \|\mathbf{h}_i\|^2} \ge \frac{K_2{}^q\|\mathbf{h}_{\nu_k}\|^2}{\sum_{i=1}^{\nu_k} \|\mathbf{h}_i\|^2 + (q-1)K_1{}^q\|\mathbf{h}_{\nu_k}\|^2}.$$

Therefore,

$$\frac{\max_{j \in J_k} \|\mathbf{h}_j\|^2 \Big/ \sum_{i=1}^{j} \|\mathbf{h}_i\|^2}{\min_{j \in J_k} \|\mathbf{h}_j\|^2 \Big/ \sum_{i=1}^{j} \|\mathbf{h}_i\|^2} \le \left(\frac{K_1}{K_2}\right)^q \left[\frac{1 + (q-1)K_1{}^q\|\mathbf{h}_{\nu_k}\|}{\sum_{j=1}^{\nu_k} \|\mathbf{h}_j\|^2}\right] \to \left(\frac{K_1}{K_2}\right)^q$$

as $k \to \infty$, by Equation 7.19*b*.

This establishes Assumption C4. Assumption C5 is immediate, since the gain vector and the gradient vector are collinear, while Assumption C6′ follows from the nonsummability of $\|\mathbf{h}_j\|^2$ and the Abel–Dini Theorem.

Since the hypotheses of Theorems 6.2 and 6.3 are identical in the linear case, we infer: *The recursion defined by the gain of Equation 7.18 generates an estimator sequence that is consistent in both the mean-square and almost-sure senses, provided that the conditions of Equations 7.17 and 7.19 are satisfied by the regression vectors* $\{\mathbf{h}_n\}$.

7.3 Optimum Gains for Recursive Linear Regression. Batch Processing

Under those circumstances where it is more natural to group scalar observations together and process them in batches, each "observation" can be thought of as a vector $\mathbf{Y}_k$ whose mean value is a vector of the form $\mathbf{H}_k'\boldsymbol{\theta}$, where $\mathbf{H}_k'$ is a matrix whose rows are the transposes of the $\mathbf{h}$-vectors that are associated with the (scalar) components of $\mathbf{Y}_k$. Thus, if $1 = \nu_1 < \nu_2 < \cdots$ and

$$\mathbf{Y}_k = \begin{bmatrix} Y_{\nu_k} \\ Y_{\nu_k+1} \\ \vdots \\ Y_{\nu_{k+1}-1} \end{bmatrix} \qquad (k = 1, 2, \cdots),$$

then

$$\mathbf{H}_k' = \begin{bmatrix} \mathbf{h}'_{\nu_k} \\ \vdots \\ \mathbf{h}'_{\nu_{k+1}-1} \end{bmatrix} \qquad (k = 1, 2, \cdots).$$

Now consider the same question that was posed in the early part of Section 7.1. If the observations $\mathbf{Y}_1, \mathbf{Y}_2, \cdots, \mathbf{Y}_{k-1}$ have been used to form the estimator $\mathbf{s}_k$, and if

$$\mathbf{R}_{k-1} = \frac{1}{\sigma^2}\mathscr{E}(\mathbf{s}_k - \boldsymbol{\theta})(\mathbf{s}_k - \boldsymbol{\theta})'$$

is known, which matrix $\mathbf{A}_k$ has the property of minimizing

$$\mathscr{E}\|\mathbf{s}_{k+1} - \boldsymbol{\theta}\|^2,$$

where

$$\mathbf{s}_{k+1} = \mathbf{s}_k + \mathbf{A}_k[\mathbf{Y}_k - \mathbf{H}_k'\mathbf{s}_k] \text{ ?} \tag{7.21}$$

By analogy with Section 7.1, we define

$$\mathbf{T}_k = (\mathbf{H}_k'\mathbf{R}_{k-1}\mathbf{H}_k + \mathbf{I}) \tag{7.22}$$

and

$$\mathbf{R}_k = \frac{1}{\sigma^2}\mathscr{E}(\mathbf{s}_{k+1} - \boldsymbol{\theta})(\mathbf{s}_{k+1} - \boldsymbol{\theta})'. \tag{7.23}$$

By substituting Equation 7.21 into Equation 7.23 and completing the square, we find that

$$\mathbf{R}_k = \mathbf{R}_{k-1} - \mathbf{R}_{k-1}\mathbf{H}_k\mathbf{T}_k^{-1}\mathbf{H}_k'\mathbf{R}_{k-1} + (\mathbf{A}_k - \mathbf{R}_{k-1}\mathbf{H}_k\mathbf{T}_k^{-1})\mathbf{T}_k(\mathbf{A}_k - \mathbf{R}_{k-1}\mathbf{H}_k\mathbf{T}_k^{-1})'. \quad (7.24)$$

The last term above is a nonnegative definite matrix whose trace is nonnegative. Therefore, $\mathscr{E}\|\mathbf{s}_{k+1} - \boldsymbol{\theta}\|^2 = \sigma^2 \operatorname{tr} \mathbf{R}_k$ is minimized when the third term vanishes, that is, when we take

$$\mathbf{A}_k = \mathbf{R}_{k-1}\mathbf{H}_k\mathbf{T}_k^{-1}.$$

In this case, we see that

$$\mathbf{R}_k = \mathbf{R}_{k-1} - \mathbf{R}_{k-1}\mathbf{H}_k\mathbf{T}_k^{-1}\mathbf{H}_k'\mathbf{R}_{k-1}.$$

If the last equation is postmultiplied by $\mathbf{H}_k$, we find that

$$\begin{aligned}\mathbf{R}_k\mathbf{H}_k &= \mathbf{R}_{k-1}\mathbf{H}_k(\mathbf{I} - \mathbf{T}_k^{-1}\mathbf{H}_k'\mathbf{R}_{k-1}\mathbf{H}_k)\\ &= \mathbf{R}_{k-1}\mathbf{H}_k[\mathbf{I} - (\mathbf{H}_k'\mathbf{R}_{k-1}\mathbf{H}_k + \mathbf{I})^{-1}\mathbf{H}_k'\mathbf{R}_{k-1}\mathbf{H}_k]\\ &= \mathbf{R}_{k-1}\mathbf{H}_k(\mathbf{H}_k'\mathbf{R}_{k-1}\mathbf{H}_k + \mathbf{I})^{-1} = \mathbf{R}_{k-1}\mathbf{H}_k\mathbf{T}_k^{-1} = \mathbf{A}_k.\end{aligned}$$

Thus, $\mathscr{E}\|\mathbf{s}_{k+1} - \boldsymbol{\theta}\|^2$ is minimized if we take

$$\mathbf{A}_k = \mathbf{R}_k\mathbf{H}_k, \quad (7.25)$$

where

$$\mathbf{R}_k = \mathbf{R}_{k-1} - \mathbf{R}_{k-1}\mathbf{H}_k(\mathbf{H}_k'\mathbf{R}_{k-1}\mathbf{H}_k + \mathbf{I})^{-1}\mathbf{H}_k'\mathbf{R}_{k-1}. \quad (7.26)$$

This, in turn, strongly suggests that we give serious consideration to recursions of the form

$$\mathbf{s}_{k+1} = \mathbf{s}_k + \mathbf{R}_k\mathbf{H}_k[\mathbf{Y}_k - \mathbf{H}_k'\mathbf{s}_k] \qquad (k = 1, 2, \cdots), \quad (7.27)$$

where $\mathbf{R}_k$ satisfies Equation 7.26 for $k = 1, 2, \cdots$ and $\mathbf{R}_0$ is arbitrary. The analogy to Section 7.1 is so close that it seems reasonable to assume that some sort of relationship exists between the recursions of Equations 7.25 through 7.27 and 7.9. This is indeed the case. In fact, the recursion of Equations 7.25 through 7.27 (suitably initialized) yields an estimator sequence which is a subsequence of the one generated by Equation 7.9.

To prove this assertion, iterate Equation 7.24 from k back to $k_0 + 1$. We find that

$$\mathbf{s}_{k+1} = \left[\prod_{j=k_0+1}^{k} (\mathbf{I} \quad \mathbf{R}_j\mathbf{H}_j\mathbf{H}_j')\right]\mathbf{s}_{k_0+1} + \sum_{j=k_0+1}^{k}\left[\prod_{i=j+1}^{k} (\mathbf{I} \quad \mathbf{R}_i\mathbf{H}_i\mathbf{H}_i')\right]\mathbf{R}_j\mathbf{H}_j\mathbf{Y}_j. \quad (7.28)$$

If $\mathbf{R}_{k_0}$ is nonsingular, it is easy to prove by induction (using Equation 7.26) that

$$\mathbf{R}_k = \left[\mathbf{R}_{k_0}^{-1} + \sum_{j=k_0+1}^{k} \mathbf{H}_j\mathbf{H}_j'\right]^{-1} \qquad (k = k_0 + 1, \cdots). \quad (7.29)$$

Thus, if $j > k_0$,

$$\mathbf{I} - \mathbf{R}_j\mathbf{H}_j\mathbf{H}_j' = \mathbf{R}_j(\mathbf{R}_j^{-1} - \mathbf{H}_j\mathbf{H}_j') = \mathbf{R}_j\mathbf{R}_{j-1}^{-1}. \quad (7.30)$$

Inserting Equation 7.30 into Equation 7.28, we see that

$$\mathbf{s}_{k+1} = \mathbf{R}_k\left(\mathbf{R}_{k_0}^{-1}\mathbf{s}_{k_0+1} + \sum_{j=k_0+1}^{k} \mathbf{H}_j\mathbf{Y}_j\right) \qquad (k \geq k_0). \quad (7.31)$$

After substituting Equation 7.29 into Equation 7.31, we obtain

$$\mathbf{s}_{k+1} = \left[\mathbf{R}_{k_0}^{-1} + \sum_{j=k_0+1}^{k} \mathbf{H}_j\mathbf{H}_j'\right]^{-1}\mathbf{R}_{k_0}^{-1}\mathbf{s}_{k_0+1} + \left[\mathbf{R}_{k_0}^{-1} + \sum_{j=k_0+1}^{k} \mathbf{H}_j\mathbf{H}_j'\right]^{-1}\left(\sum_{j=k_0+1}^{k} \mathbf{H}_j\mathbf{Y}_j\right). \quad (7.32)$$

We now consider two special initial conditions:

$$k_0 = 0, \qquad \mathbf{s}_1 = \mathbf{0}, \quad (7.33)$$

and

$$k_0 = 1, \qquad \mathbf{s}_2 = (\mathbf{H}_1\mathbf{H}_1')^{-1}\mathbf{H}_1\mathbf{Y}_1, \qquad \mathbf{R}_1 = (\mathbf{H}_1\mathbf{H}_1')^{-1}. \quad (7.34)$$

If the starting conditions are at $k_0 = 0$ according to Equation 7.33, then

$$\mathbf{s}_{k+1} = \left(\mathbf{R}_0^{-1} + \sum_{j=1}^{k} \mathbf{H}_j\mathbf{H}_j'\right)^{-1}\left(\sum_{j=1}^{k} \mathbf{H}_j\mathbf{Y}_j\right). \quad (7.35)$$

You will recall that $\mathbf{H}_j'$ is the matrix whose rows are $\mathbf{h}_{\nu_j}', \mathbf{h}_{\nu_j+1}', \cdots, \mathbf{h}_{\nu_{j+1}-1}'$, and $\mathbf{Y}_j$'s components are $Y_{\nu_j}, \cdots, Y_{\nu_{j+1}-1}$. Thus,

$$\mathbf{H}_j\mathbf{H}_j' = \sum_{i\in J_j} \mathbf{h}_i\mathbf{h}_i', \quad (7.36)$$

where $J_j = \{\nu_j, \nu_j + 1, \cdots, \nu_{j+1} - 1\}$, and

$$\mathbf{H}_j\mathbf{Y}_j = \sum_{i\in J_j} Y_i\mathbf{h}_i. \quad (7.37)$$

Substituting Equations 7.36 and 7.37 into Equation 7.35, we obtain

$$\mathbf{s}_{k+1} = \left(\mathbf{R}_0^{-1} + \sum_{j=1}^{k}\sum_{i\in J_j} \mathbf{h}_i\mathbf{h}_i'\right)^{-1}\left(\sum_{j=1}^{k}\sum_{i\in J_j} Y_i\mathbf{h}_i\right)$$

$$= \left(\mathbf{R}_0^{-1} + \sum_{j=1}^{\nu_{k+1}-1} \mathbf{h}_j\mathbf{h}_j'\right)^{-1}\left(\sum_{j=1}^{\nu_{k+1}-1} Y_j\mathbf{h}_j\right). \quad (7.38)$$

Now if we compare Equation 7.38 with Equation 7.14, we see that

$$\mathbf{s}_{k+1} = \mathbf{t}_{\nu_{k+1}}$$

if $\mathbf{R}_0 = \mathbf{B}_0$. Thus, the kth estimator generated by the batch-process recursion is the same as the ν_kth estimator generated by the scalar recursion, provided that the initial "covariance" matrices are the same.

Similarly, under the initial condition of Equation 7.34 (assuming $\mathbf{H}_1$ has rank p), we have

$$\begin{aligned}\mathbf{s}_{k+1} &= \left[\mathbf{H}_1\mathbf{H}_1' + \sum_{j=2}^{k} \mathbf{H}_j\mathbf{H}_j'\right]^{-1}\left(\sum_{j=1}^{k} \mathbf{H}_j\mathbf{Y}_j\right) \\ &= \left(\sum_{j=1}^{\nu_{k+1}-1} \mathbf{h}_j\mathbf{h}_j'\right)^{-1}\left(\sum_{j=1}^{\nu_{k+1}-1} Y_j\mathbf{h}_j\right),\end{aligned}$$

which is identical to Equation 7.15, the least-squares estimator based upon the first $\nu_{k+1} - 1$ observations.

We now recall that

$$\mathbf{B}_n = \frac{1}{\sigma^2}\,\mathscr{E}(\mathbf{t}_{n+1} - \boldsymbol{\theta})(\mathbf{t}_{n+1} - \boldsymbol{\theta})'$$

(cf. the discussion immediately preceding Equation 7.10), satisfies the recursion of Equation 7.9. Since $\mathbf{s}_k = \mathbf{t}_{\nu_k}$, it follows from Equation 7.23 that the kth element of the recursively defined sequence of Equation 7.26 is identical with the $(\nu_{k+1} - 1)$th element of the recursively defined sequence of Equation 7.7, if $k \geq k_0$ and

$$\mathbf{R}_{k_0} = \mathbf{B}_{\nu_{k_0+1}-1}.$$

In other words, the $\mathbf{R}_k$ matrices can be computed by means of the recursion of Equation 7.7, thereby circumventing the necessity of computing the matrix inversion which is called for in Equation 7.26. (Actually, the recursion (7.7) is carrying out the matrix inversion, but it is being done "painlessly.")

In summary: The batch-processing recursion,

$$\mathbf{s}_{k+1} = \mathbf{s}_k + \mathbf{A}_k[\mathbf{Y}_k - \mathbf{H}_k'\mathbf{s}_k],$$

with

$$\mathbf{A}_k = \mathbf{B}_{\nu_{k+1}-1}\mathbf{H}_k = \mathbf{B}_{\nu_{k+1}-1}(\mathbf{h}_{\nu_k}, \mathbf{h}_{\nu_k+1}, \cdots, \mathbf{h}_{\nu_{k+1}-1}) \tag{7.39}$$

and $\mathbf{B}_n$ satisfying the recursion (Equation 7.7) for $n \geq \nu_{k_0}$, generates a sequence of estimators which (depending upon initial conditions) is a subsequence of those generated by the recursion Equation 7.9, and, *consequently, their asymptotic behavior is determined by consideration of the condition of Equation 7.17.*

7.4 "Quick and Dirty" Linear Regression. Batch Processing

Preserving the notation of Sections 7.1 through 7.3, we assume that Equations 7.17 and 7.19 hold, and we consider the "batch-processing" recursion

$$\mathbf{s}_{k+1} = \mathbf{s}_k + \mathbf{A}_k^*[\mathbf{Y}_k - \mathbf{H}_k'\mathbf{s}_k],$$

where $\mathbf{A}_k^*$ is the $p \times p_k$ matrix whose column vectors are

$$\frac{\mathbf{h}_{\nu_k}}{\sum_{j=1}^{\nu_{k+1}-1} \|\mathbf{h}_j\|^2}, \quad \frac{\mathbf{h}_{\nu_k+1}}{\sum_{j=1}^{\nu_{k+1}-1} \|\mathbf{h}_j\|^2}, \quad \ldots \quad \frac{\mathbf{h}_{\nu_{k+1}-1}}{\sum_{j=1}^{\nu_{k+1}-1} \|\mathbf{h}_j\|^2}, \tag{7.40}$$

$\mathbf{Y}_k$ is the p_k-dimensional vector whose components are

$$Y_{\nu_k}, Y_{\nu_k+1}, \cdots, Y_{\nu_{k+1}-1},$$

and the ν_k are chosen to satisfy Equation 7.17.

In Section 7.2, we showed that Equations 7.17 and 7.19 imply Assumptions C1 through C6′, which are the same as Assumptions E1 through E6′ when $\mathscr{P}$ is all of Euclidean p-space. If we can show that

$$\frac{C\|\mathbf{h}_n\|}{\sum_{j=1}^{n} \|\mathbf{h}_j\|^2} \leq \frac{\|\mathbf{h}_n\|}{\sum_{j=1}^{\nu_{k+1}-1} \|\mathbf{h}_j\|^2} \leq \frac{\|\mathbf{h}_n\|}{\sum_{j=1}^{n} \|\mathbf{h}_j\|^2}$$

for $\nu_k \leq n < \nu_{k+1}$, then the arguments of Section 7.2 can be applied to the present gain vectors, and Assumptions E1 through E6′ can again be established. The inequality is indeed true. Under Equations 7.17 and 7.19, we have

$$\sum_{j=1}^{n} \|\mathbf{h}_j\|^2 \leq \sum_{j=1}^{\nu_{k+1}-1} \|\mathbf{h}_j\|^2 \leq \sum_{j=1}^{n} \|\mathbf{h}_j\|^2 + q\left[\max_{\nu_k \leq j < \nu_{k+1}} \|\mathbf{h}_j\|^2\right]$$

$$\leq \sum_{j=1}^{n} \|\mathbf{h}_j\|^2 + c_1\|\mathbf{h}_n\|^2 \leq (1 + c_1) \sum_{j=1}^{n} \|\mathbf{h}_j\|^2$$

if $\nu_k \leq n < \nu_{k+1}$. Thus, Theorem 6.4 applies to the untruncated (as well as truncated) batch-processing recursion; therefore, $\mathbf{s}_k$ converges to $\boldsymbol{\theta}$ in the mean square.

We now turn our attention to the question of gain sequences for truly nonlinear regression problems.

7.5 Gain Sequences for Recursive Nonlinear Regression. The Method of Linearization

In most applications, some prior knowledge is available concerning the true value of the regression parameter. For example, if $\boldsymbol{\theta}$ represents the vector of orbital parameters of a satellite, there is a nominal value $\boldsymbol{\theta}_0$ which is the value $\boldsymbol{\theta}$ *should* be if things were to go exactly as planned. In the absence of a major misfortune, the actual (unknown) value of $\boldsymbol{\theta}$ will be close to the nominal (known) value of $\boldsymbol{\theta}_0$. If the regression function is $F_n(\boldsymbol{\theta})$, then we can write

$$F_n(\boldsymbol{\theta}) = F_n(\boldsymbol{\theta}_0) + \dot{\mathbf{F}}_n'(\boldsymbol{\xi}_n)(\boldsymbol{\theta} - \boldsymbol{\theta}_0),$$

where $\dot{\mathbf{F}}_n(\boldsymbol{\xi}_n)$ is the gradient of F_n, evaluated at some point on the line segment which joins $\boldsymbol{\theta}$ and $\boldsymbol{\theta}_0$. Thus, if the observations Y_n are of the form

$$Y_n = F_n(\boldsymbol{\theta}) + W_n,$$

we can write

$$Y_n^* = \dot{\mathbf{F}}_n'(\boldsymbol{\xi}_n)\boldsymbol{\theta} + W_n, \tag{7.41}$$

where

$$Y_n^* = Y_n - [F_n(\boldsymbol{\theta}_0) - \dot{\mathbf{F}}_n'(\boldsymbol{\xi}_n)\boldsymbol{\theta}_0].$$

When viewed in terms of the "transformed" observations Y_n^*, Equation 7.41 looks very much like a linear-regression problem except for the fact that $\boldsymbol{\xi}_n$ (hence $\dot{\mathbf{F}}_n(\boldsymbol{\xi}_n)$ and Y_n^*) is not known. However, if $\boldsymbol{\theta}_0$ is close to $\boldsymbol{\theta}$, $\boldsymbol{\xi}_n$ must be close to $\boldsymbol{\theta}_0$ and so $\dot{\mathbf{F}}_n(\boldsymbol{\xi}_n)$ is close to $\dot{\mathbf{F}}_n(\boldsymbol{\theta}_0)$. If we let

$$\hat{Y}_n = Y_n - [F_n(\boldsymbol{\theta}_0) - \dot{\mathbf{F}}_n'(\boldsymbol{\theta}_0)\boldsymbol{\theta}_0],$$

we deduce from Equation 7.41 that

$$\hat{Y}_n \approx \dot{\mathbf{F}}_n'(\boldsymbol{\theta}_0)\boldsymbol{\theta} + W_n \tag{7.42}$$

(where $\approx$ means "approximately equal"). In turn, Equation 7.42 suggests that it would be worthwhile trying the recursive linear-regression schemes developed in Sections 7.1 through 7.4 on the transformed observations $\hat{Y}_n$. That is to say, we "pretend" that

$$\hat{Y}_n = \mathbf{h}_n'\boldsymbol{\theta} + W_n,$$

where

$$\mathbf{h}_n = \dot{\mathbf{F}}_n(\boldsymbol{\theta}_0).$$

We estimate $\boldsymbol{\theta}$ by a recursive scheme of the form

$$\mathbf{t}_{n+1} = \mathbf{t}_n + \mathbf{a}_n[\hat{Y}_n - \mathbf{h}_n'\mathbf{t}_n],$$

where $\mathbf{a}_n$ is defined in terms of $\mathbf{h}_n$ by Equations 7.9 or 7.18, and we investigate the consequences.

Actually, since we know that

$$\hat{Y}_n - \mathbf{h}_n'\mathbf{t}_n = Y_n - [F_n(\boldsymbol{\theta}_0) + \dot{\mathbf{F}}_n'(\boldsymbol{\theta}_0)(\mathbf{t}_n - \boldsymbol{\theta}_0)],$$

and since the right-hand side is approximately equal to $Y_n - F_n(\mathbf{t}_n)$ if $\mathbf{t}_n$ is near $\boldsymbol{\theta}_0$ (which it will be if $\boldsymbol{\theta}_0$ and $\mathbf{t}_n$ are both near $\boldsymbol{\theta}$), we will be equally justified in studying the recursion

$$\mathbf{t}_{n+1} = \mathbf{t}_n + \mathbf{a}_n[Y_n - F_n(\mathbf{t}_n)].$$

This is no great surprise. But by "deriving" the recursion via the technique of linearization, we have been led to consider two particularly promising gain sequences:

$$\mathbf{a}_n = \mathbf{B}_n\mathbf{h}_n, \tag{7.43}$$

where

$$\mathbf{h}_n = \dot{\mathbf{F}}_n(\boldsymbol{\theta}_0) \qquad (n = 1, 2, \cdots) \tag{7.44}$$

and

$$\mathbf{B}_n = \mathbf{B}_{n-1} - \frac{(\mathbf{B}_{n-1}\mathbf{h}_n)(\mathbf{B}_{n-1}\mathbf{h}_n)'}{1 + \mathbf{h}_n'\mathbf{B}_{n-1}\mathbf{h}_n}, \qquad (n > n_0). \tag{7.45}$$

The $\mathbf{B}_n$ recursion is initialized at n_0, where $\mathbf{B}_{n_0}$ can be any positive definite matrix. In this case, in closed form, we can write

$$\mathbf{B}_n = \left(\mathbf{B}_{n_0}^{-1} + \sum_{j=n_0+1}^{n} \mathbf{h}_j\mathbf{h}_j'\right)^{-1} \qquad \text{if} \quad n > n_0. \tag{7.46}$$

The other sequence is the nonlinear version of the "quick and dirty" gain:

$$\mathbf{a}_n = \frac{\mathbf{h}_n}{\sum_{j=1}^{n} \|\mathbf{h}_j\|^2}, \tag{7.47}$$

where $\mathbf{h}_n$ is given by Equation 7.44.

The preceding argument was based upon the idea of approximating the regression function by the first two terms of its Taylor-series expansion about some nominal parameter value $\boldsymbol{\theta}_0$, which is assumed to be close to the true parameter value. If the estimator sequence $\mathbf{t}_n$ approaches $\boldsymbol{\theta}$, then, after a time, a "better" nominal value of $\boldsymbol{\theta}$ is available. Why not, then, evaluate the gradient $\dot{\mathbf{F}}_n$ at the most current estimate for $\boldsymbol{\theta}$, and use either the gain of Equation 7.43 or 7.47, but with $\mathbf{h}_n = \dot{\mathbf{F}}_n(\mathbf{t}_n)$? Such gains are adaptive (depend upon the observations), whereas those with $\mathbf{h}_n = \dot{\mathbf{F}}_n(\boldsymbol{\theta}_0)$ are deterministic. The reader will recall from Chapter 4 that adaptive gains may or may not be more efficient in the scalar-parameter case (compare Theorem 4.2), and we feel it is safe to conjecture that a similar situation exists in the vector case.

If the data are to be processed in batches, then we consider recursions of the form

$$\mathbf{s}_{n+1} = [\mathbf{s}_n + \mathbf{A}_n(\mathbf{Y}_n - \mathbf{F}_n(\mathbf{s}_n))]_{\mathscr{P}},$$

where $\mathbf{Y}_n$ is the p_k-dimensional vector whose components are Y_{ν_n}, $Y_{\nu_n+1}, \cdots, Y_{\nu_{n+1}-1}$, $\mathbf{F}_n(\mathbf{s}_n)$ is the p_k-dimensional vector whose components are $F_{\nu_n}(\mathbf{s}_n), F_{\nu_n+1}(\mathbf{s}_n), \cdots, F_{\nu_{n+1}-1}(\mathbf{s}_n)$, and $\mathbf{A}_n$ is the $p \times p_k$ matrix whose column vectors are given either by Equation 7.39 or Equation 7.40, with $\mathbf{h}_k$ equal to the gradient of F_k evaluated either at some (fixed) nominal value $\boldsymbol{\theta}_0$ or at some recent estimate of $\boldsymbol{\theta}$.

The preceding discussion is intended to be informal and heuristic. Its purpose is to motivate the study of a few particular gain sequences in the context of our convergence theorems. In subsequent sections, we will exhibit various sufficient conditions on the regression function $F_n(\cdot)$, which imply Assumptions E1 through E6′ (E6) for the above-mentioned gains.

Before proceeding, we should point out that these gains (and their attendant recursions) have been used in practice for a long time. Their "discovery" undoubtedly occurred by means of the technique of linearization. However, to our knowledge, the question of convergence has not been treated before. For ease of future reference, we list the various gains which are to be investigated in the sequel.

7.5.1. Single-Observation Recursion. We have

$$\mathbf{t}_{n+1} = [\mathbf{t}_n + \mathbf{a}_n(\mathbf{t}_1, \mathbf{t}_2, \cdots, \mathbf{t}_n)(Y_n - F_n(\mathbf{t}_n))]_{\mathscr{P}},$$

$$\mathbf{a}_n = \frac{\dot{\mathbf{F}}_n(\boldsymbol{\xi}_n)}{\sum_{j=1}^{n} \|\dot{\mathbf{F}}_j(\boldsymbol{\xi}_j)\|^2}, \qquad (7.48a)$$

$$\mathbf{a}_n = \mathbf{B}_n \dot{\mathbf{F}}_n(\boldsymbol{\xi}_n), \qquad (7.48b)$$

where

$$\mathbf{B}_n = \left(\sum_{j=1}^{n} \dot{\mathbf{F}}_j(\boldsymbol{\xi}_j)\dot{\mathbf{F}}_j'(\boldsymbol{\xi}_j) + \mathbf{R}\right)^{-1},$$

and, for each j, $\boldsymbol{\xi}_j$ maps $(\mathbf{t}_1, \mathbf{t}_2, \cdots, \mathbf{t}_j)$ into $\mathscr{P}$. The gains can be classified as *deterministic* if $\boldsymbol{\xi}_j(\mathbf{t}_1, \mathbf{t}_2, \cdots, \mathbf{t}_j) = \boldsymbol{\theta}_0 \in \mathscr{P}$ for all j; *adaptive* if $\boldsymbol{\xi}_j(\mathbf{t}_1, \mathbf{t}_2, \cdots, \mathbf{t}_j) = \mathbf{t}_j$ $(j = 1, 2, \cdots)$; *quasi-adaptive* if $\boldsymbol{\xi}_j(\mathbf{t}_1, \mathbf{t}_2, \cdots, \mathbf{t}_j) = \mathbf{t}_{n(j)}$, where $\{n(j)\}$ is a nondecreasing sequence of integers with $n(j) \le j$.

7.5.2. Batch-Processing Recursion. Let $1 = \nu_1 < \nu_2 < \cdots$ be a sequence of integers chosen so that Assumption E3 holds. Let

$$\mathbf{s}_{k+1} = [\mathbf{s}_k + \mathbf{A}_k(\mathbf{s}_1, \cdots, \mathbf{s}_k)(\mathbf{Y}_k - \mathbf{F}_k(\mathbf{s}_k))]_{\mathscr{P}},$$

where

$$\mathbf{Y}_k = \begin{bmatrix} Y_{\nu_k} \\ \vdots \\ Y_{\nu_{k+1}-1} \end{bmatrix} \qquad \mathbf{F}_k(\cdot) = \begin{bmatrix} F_{\nu_k}(\cdot) \\ \vdots \\ F_{\nu_{k+1}-1}(\cdot) \end{bmatrix}$$

and

$$\mathbf{A}_k = (\mathbf{a}_{\nu_k}, \mathbf{a}_{\nu_k+1}, \cdots, \mathbf{a}_{\nu_{k+1}-1}).$$

Here, we consider the gains

$$\mathbf{a}_j = \frac{\dot{\mathbf{F}}_j(\boldsymbol{\xi}_j)}{\sum_{j=1}^{\nu_{k+1}-1} \|\dot{\mathbf{F}}_j(\boldsymbol{\xi}_j)\|^2} \qquad \text{if} \quad \nu_k \leq j < \nu_{k+1}, \tag{7.48c}$$

and

$$\mathbf{a}_j = \mathbf{B}_{\nu_{k+1}} \dot{\mathbf{F}}_j(\boldsymbol{\xi}_j) \qquad \text{if} \quad \nu_k \leq j < \nu_{k+1}. \tag{7.48d}$$

In the present case,

$$\boldsymbol{\xi}_{\nu_k} = \boldsymbol{\xi}_{\nu_k+1} = \cdots = \boldsymbol{\xi}_{\nu_{k+1}-1}$$

map $(\mathbf{s}_1, \cdots, \mathbf{s}_k)$ into $\mathscr{P}$. The gains can be classified as

$$\left.\begin{array}{ll} \textit{Deterministic} \text{ if } \boldsymbol{\xi}_{\nu_k} & = \boldsymbol{\theta}_0 \in \mathscr{P} \\ \textit{Adaptive} \text{ if } \boldsymbol{\xi}_{\nu_k} & = \mathbf{s}_k \\ \textit{Quasi-adaptive} \text{ if } \boldsymbol{\xi}_{\nu_k} & = \mathbf{s}_{n(k)} \end{array}\right\} \quad (k = 1, 2, \cdots),$$

where $n(k)$ is a nondecreasing integer sequence with $n(k) \leq k$.

7.6 Sufficient Conditions for Assumptions E1 Through E6′ (E6) When the Gains (Equations 7.48) Are Used

THEOREM 7.1

Let $\{F_n(\cdot)\}$ be a sequence of real-valued functions defined over a p-dimensional closed convex set $\mathscr{P}$. We assume that each $F_n(\cdot)$ has bounded, second-order mixed partial derivatives over $\mathscr{P}$ and that for some $\mathbf{x} \in \mathscr{P}$, the following conditions hold true.

F1. $\limsup_{\mathbf{y} \in \mathscr{P}^{(p)}} \|\mathbf{G}_n(\mathbf{y})\| / \|\mathbf{h}_n^*\| < \infty$,

where

$$\mathbf{G}_n(\mathbf{y}) = \mathbf{G}_n(\mathbf{y}_1, \mathbf{y}_2, \cdots, \mathbf{y}_p)$$

is the $p \times p$ matrix whose ith column is

$$\begin{bmatrix} \partial^2 F_n(\xi)/\partial\xi_1\partial\xi_i \\ \partial^2 F_n(\xi)/\partial\xi_2\partial\xi_i \\ \vdots \\ \partial^2 F_n(\xi)/\partial\xi_p\partial\xi_i \end{bmatrix}_{\xi = \mathbf{y}_i} \qquad (i = 1, 2, \cdots, p)$$

and

$$\mathbf{h}_n^* = \dot{\mathbf{F}}_n(\mathbf{x}) \qquad \text{satisfies the conditions:}$$

F2. $\lim_{n\to\infty} \|\mathbf{h}_n^*\|^2 / \sum_{j=1}^{n} \|\mathbf{h}_j^*\|^2 = 0.$

F3. $\sum_{n=1}^{\infty} \|\mathbf{h}_n^*\|^2 = \infty.$

F4. $\limsup_{n\to\infty} \max(\|\mathbf{h}_n^*\|/\|\mathbf{h}_{n+1}^*\|, \|\mathbf{h}_{n+1}^*\|/\|\mathbf{h}_n^*\|) = K < \infty.$

F5. **There is a sequence of integers**

$$1 = \nu_1 < \nu_2 < \cdots,$$

with

$$p \le \nu_{k+1} - \nu_k = p_k \le q < \infty,$$

such that

$$\liminf_{k\to\infty} \frac{1}{p_k} \lambda_{\min}\left(\sum_{j\in J_k} \frac{\mathbf{h}_j^* \mathbf{h}_j^{*\prime}}{\|\mathbf{h}_j^*\|^2}\right) = \tau^{*2} > 0,$$

where

$$J_k = \{\nu_k, \nu_k + 1, \cdots, \nu_{k+1} - 1\}.$$

Let $r(\mathscr{P})$ be the radius of the smallest closed sphere containing $\mathscr{P}$.

a. For the gains of Equations 7.48*a*, *c*, Assumptions E1 through E6′ hold if $r(\mathscr{P})$ is sufficiently small. (In the case of batch processing, we intend that the batches of data correspond to the index sets induced by Assumption F5. Thus, the kth batch of data consists of $\{Y_j; j \in J_k\}$.)

b. If Assumptions F2, F3, and F4 are strengthened to

F2′. $K_1 n^\delta \le \|\mathbf{h}_n^*\| \le K_2 n^\delta$ for some positive K_1, K_2, and δ, then Assumptions E1 through E6 hold for the gains of Equations 7.48*a*, *c* if $r(\mathscr{P})$ is sufficiently small.

c. For the gains of Equations 7.48*b*, *d*, Assumptions E1 through E6′ hold if we assume, in addition to Assumptions F1 through F5, that

F6. $$\frac{4(\tau^*/K^q\sigma^*)^2}{[1+(\tau^*/K^q\sigma^*)^2]^2} > \frac{1-(p/q)\tau^{*2}}{1-(p/q)\tau^{*2}+(1/pqK^{8q})\tau^{*6}},$$

provided $r(\mathscr{P})$ is sufficiently small, where

$$\sigma^{*2} = \limsup_{k\to\infty} \frac{1}{p_k} \lambda_{\max}\left(\sum_{j\in J_k} \frac{\mathbf{h}_j^*\mathbf{h}_j^{*\prime}}{\|\mathbf{h}_j^*\|^2}\right).$$

d. For the gains of Equations 7.48*b*, *d*, Assumptions F1, F2', F5, and F6 together imply Assumptions E1 through E6, provided that $r(\mathscr{P})$ is suitably small.

Comment: In practice, the true parameter value $\boldsymbol{\theta}$ is not known exactly but is generally known to lie in some neighborhood $\mathscr{P}$ of a nominal value $\boldsymbol{\theta}_0$. In such cases, the vectors $\mathbf{h}_n^*$ would most naturally be chosen equal to $\dot{\mathbf{F}}_n(\boldsymbol{\theta}_0)$. Theorem 7.1 says that Assumptions E1 through E6' will obtain under various subsets of Assumptions F1 through F6 if $\mathscr{P}$ is "sufficiently small." Just *how* small could be specified quantitatively, as will be seen in the proof. However, since a quantitative bound is so complicated and conservative that we feel it contributes little to our understanding, we do not include it. The purpose of Theorem 7.1 is to furnish a set of relatively easy to understand conditions that furnish insight into the circumstances under which our estimator recursions will converge, always subject to the proviso that $r(\mathscr{P})$ should fall below some threshold value.

Proof of a. If $\boldsymbol{\xi}\in\mathscr{P}$, we can write $\dot{\mathbf{F}}_n(\boldsymbol{\xi}) = \mathbf{h}_n^* + \mathbf{G}_n(\boldsymbol{\xi}-\mathbf{x})$, where $\mathbf{G}_n$ is the matrix of $\dot{\mathbf{F}}_n$'s second-order mixed partials evaluated at various points in $\mathscr{P}$. If we let

$$\mathbf{r}_n(\boldsymbol{\xi}) = \dot{\mathbf{F}}_n(\boldsymbol{\xi}) - \mathbf{h}_n^*, \tag{7.49a}$$

it follows from Assumption F1 that

$$\sup_{\boldsymbol{\xi}\in\mathscr{P}} \|\mathbf{r}_n(\boldsymbol{\xi})\| \le C_1 r(\mathscr{P})\|\mathbf{h}_n^*\|. \tag{7.49b}$$

Since

$$\|\mathbf{r}_n(\boldsymbol{\xi})\| \ge \max(\|\dot{\mathbf{F}}_n(\boldsymbol{\xi})\| - \|\mathbf{h}_n^*\|, \|\mathbf{h}_n^*\| - \|\dot{\mathbf{F}}_n(\boldsymbol{\xi})\|),$$

it therefore follows that

$$0 < (1 - C_1 r(\mathscr{P}))\|\mathbf{h}_n^*\| \le \|\dot{\mathbf{F}}_n(\boldsymbol{\xi})\| \le (1 + C_1 r(\mathscr{P}))\|\mathbf{h}_n^*\| \tag{7.50}$$

for all $\boldsymbol{\xi}\in\mathscr{P}$ if $r(\mathscr{P})$ is chosen small enough to ensure the leftmost inequality. By Assumption F4, we have

$$\limsup_{k\to\infty} \max_{i,j\in J_k} \|\mathbf{h}_i^*\|/\|\mathbf{h}_j^*\| \le K^q; \tag{7.51}$$

therefore, for the gains of Equations 7.48*a*, *c*, there exist positive constants C_2 and C_2' such that

$$C_2(1 - C_1 r(\mathscr{P}))\|\mathbf{h}_n^*\| \le \|\mathbf{a}_n\| \sum_{j=1}^{n} \|\mathbf{h}_j^*\|^2 \le C_2'(1 + C_1 r(\mathscr{P}))\|\mathbf{h}_n^*\| \tag{7.52}$$

uniformly in $\mathbf{a}_n$'s argument for all n. Assumptions E1, E2, E4, and E6′ now follow when Equations 7.49 through 7.52, F2, F3, and the Abel–Dini theorem (2.27) are combined in what is, by now, routine fashion.

To prove Assumption E5 for the gains of Equations 7.48*a*, *c*, we notice that

$$\frac{\mathbf{a}_j}{\|\mathbf{a}_j\|} = \frac{\dot{\mathbf{F}}_j(\xi_j)}{\|\dot{\mathbf{F}}_j(\xi_j)\|};$$

therefore, by Equation 7.49*a*, we have

$$\frac{\mathbf{a}_j'\dot{\mathbf{F}}_j(\mathbf{y})}{\|\mathbf{a}_j\|\ \|\dot{\mathbf{F}}_j(\mathbf{y})\|} = \frac{\mathbf{h}_j^{*\prime}\mathbf{h}_j^* + \mathbf{r}_j'(\mathbf{y})\mathbf{r}_j(\xi_j) + (\mathbf{r}_j'(\mathbf{y}) + \mathbf{r}_j'(\xi_j))\mathbf{h}_j^*}{\|\mathbf{h}_j^* + \mathbf{r}_j(\xi_j)\|\cdot\|\mathbf{h}_j^* + r_j(\mathbf{y})\|}. \tag{7.53}$$

Multiplying the numerator and denominator of the last expression by $\|\mathbf{h}_j^*\|^2$, we find that

$$\frac{\mathbf{a}_j'\dot{\mathbf{F}}_j(\mathbf{y})}{\|\mathbf{a}_j\|\ \|\dot{\mathbf{F}}_j(\mathbf{y})\|} = \left[\frac{\|\mathbf{h}_j^*\|}{\|\mathbf{h}_j^* + \mathbf{r}_j(\xi_j)\|}\right]\left[\frac{\|\mathbf{h}_j^*\|}{\|\mathbf{h}_j^* + \mathbf{r}_j(\mathbf{y})\|}\right] \times \left[1 + \frac{\mathbf{r}_j'(\mathbf{y})\mathbf{r}_j(\xi_j) + (\mathbf{r}_j(\mathbf{y}) + \mathbf{r}_j(\xi_j))'\mathbf{h}_j^*}{\|\mathbf{h}_j^*\|^2}\right].$$

Using Schwarz's Inequality and Inequality 7.49*b*, we see that

$$1 + \frac{\mathbf{r}_j'(\mathbf{y})\mathbf{r}_j(\xi_j) + (\mathbf{r}_j(\mathbf{y}) + \mathbf{r}_j(\xi_j))'\mathbf{h}_j^*}{\|\mathbf{h}_j^*\|^2} \ge 1 - C_3 r(\mathscr{P})$$

and, by Equation 7.49*b*, that

$$\frac{\|\mathbf{h}_j^*\|}{\|\mathbf{h}_j^* + \mathbf{r}_j(\xi)\|} \ge 1 - C_4 r(\mathscr{P})$$

for all $\xi \in \mathscr{P}$, provided that $r(\mathscr{P})$ is suitably small. Thus, the left-hand side of Equation 7.53 is bounded below (uniformly) by $(1 - C_3 r(\mathscr{P})) \times (1 - C_4 r(\mathscr{P}))^2$, which can be made arbitrarily close to one by taking $r(\mathscr{P})$ small enough. Since Assumption E5 requires the left-hand side of Equation 7.53 to be bounded below uniformly by some number which is strictly less than one, Assumption E5 will therefore hold if $r(\mathscr{P})$ is suitably small.

Assumption E3 does not depend upon the particular gain sequence, and so we prove it now, once and for all, as follows:

$$\sum_{j \in J_k} \frac{\dot{\mathbf{F}}_j(\mathbf{x}_j)\dot{\mathbf{F}}_j'(\mathbf{x}_j)}{\|\dot{\mathbf{F}}_j(\mathbf{x}_j)\|^2}$$
$$= \sum_{j \in J_k} \frac{\mathbf{h}_j^*\mathbf{h}_j^{*\prime}}{\|\mathbf{h}_j^*\|^2} \frac{\|\mathbf{h}_j^*\|^2}{\|\mathbf{h}_j^* + \mathbf{r}_j\|^2} + \sum_{j \in J_k} \frac{\mathbf{r}_j\mathbf{r}_j'}{\|\mathbf{h}_j^* + \mathbf{r}_j\|^2} + \sum_{j \in J_k} \frac{\mathbf{r}_j\mathbf{h}_j^{*\prime} + \mathbf{h}_j^*\mathbf{r}_j'}{\|\mathbf{h}_j^* + \mathbf{r}_j\|^2}$$
$$= 1^\circ + 2^\circ + 3^\circ, \tag{7.54}$$

where

$$\mathbf{r}_j = \dot{\mathbf{F}}_j(\mathbf{x}_j) - \mathbf{h}_j^*.$$

By Equation 7.49*b*, if $r(\mathscr{P})$ is sufficiently small, we have

$$\frac{\|\mathbf{h}_j^*\|}{\|\mathbf{h}_j^* + \mathbf{r}_j\|} = 1 + \gamma_j, \tag{7.55}$$

where $|\gamma_j| \le C_5 r(\mathscr{P})$. So, if $\varepsilon > 0$ is given,

$$\lambda_{\min}(1^\circ) \ge \lambda_{\min}\left(\sum_{j \in J_k} \frac{\mathbf{h}_j^*\mathbf{h}_j^{*\prime}}{\|\mathbf{h}_j^*\|^2}\right) - \operatorname{tr} \sum_{j \in J_k} |\gamma_j| \frac{\mathbf{h}_j^*\mathbf{h}_j^{*\prime}}{\|\mathbf{h}_j^*\|^2}$$
$$\ge p(\tau^{*2} - \varepsilon) - qC_5 r(\mathscr{P})$$

by Assumption F5 if k is suitably large and $r(\mathscr{P})$ is suitably small. Here 2° is a nonnegative definite matrix. From the Courant–Fischer characterization of eigenvalues, Schwarz's Inequality, and Equations 7.49*b* and 7.55),

$$\lambda_{\min}(3^\circ) \ge -2 \sum_{j \in J_k} \frac{\|\mathbf{r}_j\|\ \|\mathbf{h}_j^*\|}{\|\mathbf{r}_j + \mathbf{h}_j^*\|^2}$$
$$= -2 \sum_{j \in J_k} \frac{\|\mathbf{r}_j\|}{\|\mathbf{h}_j^*\|} \left(\frac{\|\mathbf{h}_j^*\|}{\|\mathbf{r}_j + \mathbf{h}_j^*\|}\right)^2 \ge -C_6 r(\mathscr{P}).$$

Hence, by Equation 7.54,

$$\tau^2 = \liminf_{k \to \infty} \frac{1}{p_k} \inf \lambda_{\min}\left(\sum_{j \in J_k} \frac{\dot{\mathbf{F}}_j(\mathbf{x}_j)\dot{\mathbf{F}}_j'(\mathbf{x}_j)}{\|\dot{\mathbf{F}}_j(\mathbf{x}_j)\|^2}\right) \ge \frac{p(\tau^{*2} - \varepsilon) - C_7 r(\mathscr{P})}{q}$$

(where the infimum is taken as the $\mathbf{x}_j$'s vary over p-space).

Since ε is arbitrary, we see that

$$\tau^2 > 0, \tag{7.56}$$

provided that $r(\mathscr{P})$ is appropriately small. This proves E3.

Proof of b. Since Assumption F2′ implies Assumptions F2 through F4, Assumptions El through E5 hold as before and Equation 7.50 can be strengthened to

$$K_1(1 - C_1 r(\mathscr{P}))n^{\delta} \leq \|\dot{\mathbf{F}}_n(\xi)\| \leq K_2(1 + C_1 r(\mathscr{P}))n^{\delta},$$

for all $\xi \in \mathscr{P}$. Thus,

$$\|\mathbf{a}_n\| \leq \frac{K_3 n^{\delta}}{\sum_{j=1}^{n} j^{2\delta}} = O\left(\frac{1}{n^{1+\delta}}\right)$$

which implies Assumption E6.

Proof of c. For the gains of Equations 7.48*b*, *d*, we have

$$\lambda_{\min}(\mathbf{B}_{\nu_{k+1}-1})\|\dot{\mathbf{F}}_n(\xi_n)\| \leq \|\mathbf{a}_n\| \leq \lambda_{\max}(\mathbf{B}_n)\|\dot{\mathbf{F}}_n(\xi_n)\|, \tag{7.57}$$

where

$$\mathbf{B}_m = \left(\sum_{j=1}^{m} \dot{\mathbf{F}}_j(\xi_j)\dot{\mathbf{F}}_j'(\xi_j) + \mathbf{R}\right)^{-1} \qquad (m = n, \nu_{k+1} - 1),$$

and k is chosen so that

$$\nu_k \leq n < \nu_{k+1}.$$

By Equation 7.49*a*, we see that

$$\left[\lambda_{\max}\left(\sum_{j=1}^{\nu_{k+1}-1} (\mathbf{h}_j^* + \mathbf{r}_j)(\mathbf{h}_j^* + \mathbf{r}_j)' + \mathbf{R}\right)\right]^{-1} \big|\|\mathbf{h}_n^*\| - \|\mathbf{r}_n\|\big| \leq \|\mathbf{a}_n\|$$

$$\leq \left[\lambda_{\min}\left(\sum_{j=1}^{n} (\mathbf{h}_j^* + \mathbf{r}_j)(\mathbf{h}_j^* + \mathbf{r}_j)' + \mathbf{R}\right)\right]^{-1} (\|\mathbf{h}_n^*\| + \|\mathbf{r}_n\|). \tag{7.58}$$

By Equation 7.49*b*, we have

$$\|\mathbf{h}_n^*\| - \|\mathbf{r}_n\| \geq (1 - C_1 r(\mathscr{P}))\|\mathbf{h}_n^*\|, \quad \text{and} \quad \|\mathbf{h}_n^*\| + \|\mathbf{r}_n\| \leq (1 + C_1 r(\mathscr{P}))\|\mathbf{h}_n^*\|. \tag{7.59}$$

Furthermore, since $\mathbf{R}$ is assumed to be nonnegative definite and since $\lambda_{\min}(\mathbf{A} + \mathbf{B}) \geq \lambda_{\min}(\mathbf{A}) + \lambda_{\min}(\mathbf{B})$ for symmetric matrices,

$$\lambda_{\min}\left(\sum_{j=1}^{n} (\mathbf{h}_j^* + \mathbf{r}_j)(\mathbf{h}_j^* + \mathbf{r}_j)' + \mathbf{R}\right)$$

$$\geq \lambda_{\min}\left(\sum_{j=1}^{n} \mathbf{h}_j^*\mathbf{h}_j^{*\prime}\right) - 2\sum_{j=1}^{n} \|\mathbf{r}_j\| \cdot \|\mathbf{h}_j^*\|$$

$$\geq \lambda_{\min}\left(\sum_{j=1}^{n} \mathbf{h}_j^*\mathbf{h}_j^{*\prime}\right) - 2C_1 r(\mathscr{P}) \sum_{j=1}^{n} \|\mathbf{h}_j^*\|^2.$$

By Lemma 7*b*, if $\varepsilon > 0$ is given,

$$\lambda_{\min}\left(\sum_{j=1}^{n} \mathbf{h}_j^*\mathbf{h}_j^{*\prime}\right) \geq \left(\frac{\tau^{*2} - \varepsilon}{K^{2q}}\right) \lambda_{\max}\left(\sum_{j=1}^{n} \mathbf{h}_j^*\mathbf{h}_j^{*\prime}\right)$$

$$\geq \frac{1}{p}\left(\frac{\tau^{*2} - \varepsilon}{K^{2q}}\right) \sum_{j=1}^{n} \|\mathbf{h}_j^*\|^2$$

if n is large.

Consequently,

$$\lambda_{\min}\left(\sum_{j=1}^{n} (\mathbf{h}_j^* + \mathbf{r}_j)(\mathbf{h}_j^* + \mathbf{r}_j)' + \mathbf{R}\right)$$

$$\geq \left[\left(\frac{\tau^{*2} - 2\varepsilon}{pK^{2q}}\right) - 2C_1 r(\mathscr{P})\right] \sum_{j=1}^{n} \|\mathbf{h}_j^*\|^2. \quad (7.60)$$

Combining Equations 7.57, 7.50, and 7.60, we obtain

$$\|\mathbf{a}_n\| \leq \left[\left(\frac{\tau^{*2} - 2\varepsilon}{pK^{2q}}\right) - 2C_1 r(\mathscr{P})\right]^{-1} (1 + C_1 r(\mathscr{P})) \frac{\|\mathbf{h}_n^*\|}{\sum_{j=1}^{n} \|\mathbf{h}_j^*\|^2}, \quad (7.61)$$

when n is large (the result holding uniformly in $\mathbf{a}_n$'s argument, of course). By Assumption F3, Equations 7.50, 7.61, and the Abel–Dini Theorem (2.27), Assumption E6′ holds, while Equations 7.50, 7.61, and Assumption F2 imply Assumption E1.

On the other hand, since $n \in J_k$, we have

$$\lambda_{\max}\left(\sum_{j=1}^{\nu_{k+1}-1} (\mathbf{h}_j^* + \mathbf{r}_j)(\mathbf{h}_j^* + \mathbf{r}_j)' + \mathbf{R}\right)$$

$$\leq \sum_{j=1}^{\nu_{k+1}-1} \|\mathbf{h}_j^* + \mathbf{r}_j\|^2 + \operatorname{tr} \mathbf{R}$$

$$\leq \left(\sum_{j=1}^{n} + \sum_{j \in J_k}\right) (\|\mathbf{h}_j^*\|^2 + 2\|\mathbf{r}_j\|\,\|\mathbf{h}_j^*\| + \|\mathbf{r}_j\|^2) + \operatorname{tr} \mathbf{R}$$

$$\leq (1 + C_1 r(\mathscr{P}))^2 \left(\sum_{j=1}^{n} \|\mathbf{h}_j^*\|^2 + q \max_{j \in J_k} \|\mathbf{h}_j^*\|^2\right) + \operatorname{tr} \mathbf{R}. \quad (7.62)$$

By Equation 7.51,

$$\max_{j \in J_k} \|\mathbf{h}_j^*\|^2 \leq K^{2q}\|\mathbf{h}_n^*\|^2 \qquad \text{if} \quad n \in J_k,$$

and by Assumptions F2 and F3, if $\varepsilon > 0$ is given, we have

$$\frac{\operatorname{tr} \mathbf{R} + q(1 + C_1 r(\mathscr{P}))^2 K^{2q}\|\mathbf{h}_n^*\|^2}{\sum_{j=1}^{n} \|\mathbf{h}_j^*\|^2} < \varepsilon$$

for large n. Combining Equations 7.58, 7.59, and 7.62, we find that

$$\|\mathbf{a}_n\| \geq [(1 + C_1 r(\mathscr{P}))^2 + \varepsilon]^{-1}(1 - C_1 r(\mathscr{P})) \left(\frac{\|\mathbf{h}_n^*\|}{\sum_{j=1}^{n} \|\mathbf{h}_j^*\|^2}\right) \tag{7.63}$$

if n is large. Equation 7.63 and the Abel–Dini Theorem imply Assumption E2. To prove Assumption E4, we notice that, for $n \in J_k$,

$$\frac{\min_{j \in J_k} \|\mathbf{h}_j^*\|^2}{\sum_{j=1}^{\nu_k} \|\mathbf{h}_j^*\|^2 + q \max_{j \in J_k} \|\mathbf{h}_j^*\|^2} \leq \frac{\|\mathbf{h}_n^*\|^2}{\sum_{j=1}^{n} \|\mathbf{h}_j^*\|^2} \leq \frac{\max_{j \in J_k} \|\mathbf{h}_j^*\|^2}{\sum_{j=1}^{\nu_k} \|\mathbf{h}_j^*\|^2}. \tag{7.64}$$

Thus, by Equations 7.61, 7.63, 7.64, and 7.50,

$$\frac{\max_{n \in J_k} \|\mathbf{a}_n\| \, \|\dot{\mathbf{F}}_n\|}{\min_{n \in J_k} \|\mathbf{a}_n\| \, \|\dot{\mathbf{F}}_n\|} \leq \left[\frac{1 + C_1 r(\mathscr{P})}{1 - C_1 r(\mathscr{P})}\right]^2 \left[\frac{[(1 + C_1 r(\mathscr{P}))^2 + \varepsilon](pK^{2q})}{\tau^{*2} - 2\varepsilon - 2pK^{2q} C_1 r(\mathscr{P})}\right]$$
$$\times \left[\frac{\max_{j \in J_k} \|\mathbf{h}_j^*\|^2}{\min_{j \in J_k} \|\mathbf{h}_j^*\|^2}\right] \left[1 + q \frac{\max_{j \in J_k} \|\mathbf{h}_j^*\|^2}{\sum_{j=1}^{\nu_k} \|\mathbf{h}_j^*\|^2}\right]$$

By Equation 7.51, we see that

$$\frac{\max_{j \in J_k} \|\mathbf{h}_j^*\|^2}{\min_{j \in J_k} \|\mathbf{h}_j^*\|^2} \leq K^{2q}, \qquad \max_{j \in J_k} \|\mathbf{h}_j^*\|^2 \leq \|\mathbf{h}_{\nu_k}^*\|^2 K^{2q},$$

and, by virtue of Assumption F2,

$$\frac{\|\mathbf{h}_{\nu_k}^*\|^2}{\sum_{j=1}^{\nu_k} \|\mathbf{h}_j^*\|^2} \leq \frac{\varepsilon}{qK^{2q}}$$

if k is large (that is, if n is large). Thus, ρ (as defined by Assumption E4) is bounded above by

$$\left(\frac{1 + C_1 r(\mathscr{P})}{1 - C_1 r(\mathscr{P})}\right)^2 (pK^{4q}) \left[\frac{(1 - C_1 r(\mathscr{P}))^2 + \varepsilon}{\tau^{*2} - 2pK^{2q} C_1 r(\mathscr{P}) - 2\varepsilon}\right] (1 + \varepsilon).$$

Since ε is arbitrary,

$$\rho \leq \frac{(1 + C_1 r(\mathscr{P}))^4 (pK^{4q})}{(1 - C_1 r(\mathscr{P}))^2 (\tau^{*2} - 2pK^{2q} C_1 r(\mathscr{P}))}. \tag{7.65}$$

This establishes Assumption E4 and will play a role in the proof of Assumption E5.

By Equation 7.49*a*, we can always write

$$\dot{\mathbf{F}}_n(\mathbf{y}) = \dot{\mathbf{F}}_n(\xi_n) + \mathbf{r}_n^*(\xi_n, \mathbf{y}),$$

where

$$\mathbf{r}_n^*(\xi, \mathbf{y}) = \mathbf{r}_n(\mathbf{y}) - \mathbf{r}_n(\xi)$$

satisfies

$$\|\mathbf{r}_n^*\| \leq 2C_1 \mathbf{r}(\mathscr{P})\|\mathbf{h}_n^*\|$$

for all ξ, $\mathbf{y} \in \mathscr{P}$.

Thus, for the gains of Equations 7.48*b*, *d*,

$$\frac{\mathbf{a}_n'\dot{\mathbf{F}}_n(\mathbf{y})}{\|\mathbf{a}_n\|\ \|\dot{\mathbf{F}}_n(\mathbf{y})\|} = \frac{\dot{\mathbf{F}}_n'(\xi_n)\mathbf{B}_m\dot{\mathbf{F}}_n(\xi_n) + \dot{\mathbf{F}}_n'(\xi_n)\mathbf{B}_m\mathbf{r}_n^*}{\|\mathbf{B}_m\dot{\mathbf{F}}_n(\xi_n)\| \cdot \|\dot{\mathbf{F}}_n(\xi_n) + \mathbf{r}_n^*\|}$$

$$\geq \left[\frac{\dot{\mathbf{F}}_n'(\xi_n)\mathbf{B}_m\dot{\mathbf{F}}_n(\xi_n)}{\|\mathbf{B}_m\dot{\mathbf{F}}_n(\xi_n)\| \cdot \|\dot{\mathbf{F}}_n(\xi_n)\|}\right]\left[\frac{\|\dot{\mathbf{F}}_n(\xi_n)\|}{\|\dot{\mathbf{F}}_n(\xi_n)\| + \|\mathbf{r}_n^*\|}\right] - \frac{\|\mathbf{r}_n^*\|\ \|\mathbf{B}_m\dot{\mathbf{F}}_n(\xi_n)\|}{\|\mathbf{B}_m\dot{\mathbf{F}}_n(\xi_n)\| \cdot \left|\|\dot{\mathbf{F}}_n(\xi_n)\| - \|\mathbf{r}_n^*\|\right|},$$

where

$$m = \begin{cases} n, & \text{for Equation 7.48}b, \\ \nu_{k+1} - 1, & \text{for Equation 7.48}d. \end{cases}$$

By Equations 7.50 and 7.66,

$$\|\dot{\mathbf{F}}_n(\xi_n)\| + \|\mathbf{r}_n^*\| \leq (1 + 3C_1 r(\mathscr{P}))\|\mathbf{h}_n^*\|,$$

$$\|\dot{\mathbf{F}}_n(\xi_n)\| \geq (1 - C_1 r(\mathscr{P}))\|\mathbf{h}_n^*\|\ ,$$

and

$$\left|\|\dot{\mathbf{F}}_n(\xi_n)\| - \|\mathbf{r}_n^*\|\right| \geq (1 - 3C_1 r(\mathscr{P}))\|\mathbf{h}_n^*\|.$$

Letting

$$\kappa_m = \frac{\lambda_{\min}(\mathbf{B}^{-1})}{\lambda_{\max}(\mathbf{B}^{-1})},$$

by Lemma 7*a* we have,

$$\frac{\dot{\mathbf{F}}_n'(\xi_n)\mathbf{B}_m\dot{\mathbf{F}}_n(\xi_n)}{\|\mathbf{B}_m\dot{\mathbf{F}}_n(\xi_n)\|\ \|\dot{\mathbf{F}}_n(\xi_n)\|} \geq 2\kappa_m^{1/2}(1 + \kappa_m)^{-1}.$$

Thus, we see that

$$\liminf_n \inf_{\mathbf{x} \in \mathscr{P}(n)\mathbf{y} \in \mathscr{P}} \frac{\mathbf{a}_n'(\mathbf{x})\dot{\mathbf{F}}_n(\mathbf{y})}{\|\mathbf{a}_n(\mathbf{x})\|\ \|\dot{\mathbf{F}}_n(\mathbf{y})\|}$$

$$\geq \liminf_m 2\kappa_m^{1/2}(1 + \kappa_m)^{-1}\left(\frac{1 - 2C_1 r(\mathscr{P})}{1 + 3C_1 r(\mathscr{P})}\right) - \frac{2C_1 r(\mathscr{P})}{1 - 3C_1 r(\mathscr{P})}$$

$$\geq \liminf_m 2\kappa_m^{1/2}(1 + \kappa_m)^{-1} - C_8 r(\mathscr{P}) \tag{7.66}$$

if $r(\mathscr{P})$ is small.

If we can show that

$$\liminf_{n\to\infty} \kappa_n \geq \left(\frac{\tau^*}{K^q\sigma^*}\right)^2 - C_9 r(\mathscr{P}), \tag{7.67}$$

then E5 will follow if $r(\mathscr{P})$ is small enough. This is so because

$$\tau^2 \geq \left(\frac{p}{q}\right)\tau^{*2} - C_{10}r(\mathscr{P}), \qquad \text{and} \qquad \rho \leq \left(\frac{pK^{4q}}{\tau^{*2}}\right) + C_{11}r(\mathscr{P}),$$

when $r(\mathscr{P})$ is small, by virtue of Equations 7.56 and 7.65. Since $(1 - \tau^2)/(1 - \tau^2 + \tau^2/\rho^2)$ decreases with τ^2 and increases with ρ^2, it follows that

$$\frac{1 - \tau^2}{1 - \tau^2 + \tau^2/\rho^2} \leq \frac{1 - (p/q)\tau^{*2}}{1 - (p/q)\tau^{*2} + (1/pqK^{8q})\tau^{*6}} + C_{12}r(\mathscr{P})$$

if $r(\mathscr{P})$ is small.

On the other hand, if Equation 7.67 holds,

$$\liminf_{n\to\infty} 2\kappa_n^{1/2}(1 + \kappa_n)^{-1} \geq \left(\frac{2\tau^*}{K^q\sigma^*}\right)\left[1 + \left(\frac{\tau^*}{K^q\sigma^*}\right)^2\right]^{-1} - C_{13}r(\mathscr{P}),$$

and so, by Assumption F6,

$$\liminf_{n\to\infty} 2\kappa_n^{1/2}(1 + \kappa_n)^{-1} > \left(\frac{1 - \tau^2}{1 - \tau^2 + \tau^2/\rho^2}\right)^{1/2}$$

if $r(\mathscr{P})$ is sufficiently small. Thus, E5 follows from Equation 7.66, when $r(\mathscr{P})$ is small.

To prove Equation 7.67, we note that

$$\lambda_{\min}(\mathbf{B}_m^{-1}) \geq \lambda_{\min}(\mathbf{B}_{\nu_k-1}^{-1})$$
$$\geq \sum_{n=1}^{k-1} \lambda_{\min}\left[\sum_{j\in J_n} (\mathbf{h}_j^* + \mathbf{r}_j)(\mathbf{h}_j^* + \mathbf{r}_j)'\right] \qquad \text{if} \quad m \in J_k.$$

By virtue of the fact that

$$\lambda_{\min}(\mathbf{B}_m^{-1}) = \min_{\|\mathbf{x}\|=1} \mathbf{x}'\mathbf{B}_m^{-1}\mathbf{x},$$

we see that

$$\lambda_{\min}\left[\sum_{j\in J_n} (\mathbf{h}_j^* + \mathbf{r}_j)(\mathbf{h}_j^* + \mathbf{r}_j)'\right]$$
$$\geq \min_{j\in J_n} \|\mathbf{h}_j^*\|^2 \lambda_{\min}\left[\sum_{j\in J_n} \frac{(\mathbf{h}_j^* + \mathbf{r}_j)(\mathbf{h}_j^* + \mathbf{r}_j)'}{\|\mathbf{h}_j^*\|^2}\right]$$
$$\geq \min_{j\in J_n} \|\mathbf{h}_j^*\|^2 \left\{\lambda_{\min}\left[\sum_{j\in J_n} \frac{\mathbf{h}_j^*\mathbf{h}_j^{*\prime}}{\|\mathbf{h}_j^*\|^2}\right] + \lambda_{\min}\left[\sum_{j\in J_n} \frac{\mathbf{r}_j\mathbf{h}_j^{*\prime} + \mathbf{h}_j^*\mathbf{r}_j'}{\|\mathbf{h}_j^*\|^2}\right]\right\}$$
$$\geq \min_{j\in J_n} \|\mathbf{h}_j^*\|^2 \left\{\lambda_{\min}\left[\sum_{j\in J_n} \frac{\mathbf{h}_j^*\mathbf{h}_j^{*\prime}}{\|\mathbf{h}_j^*\|^2}\right] - 2\sum_{j\in J_n} \frac{\|\mathbf{r}_j\|}{\|\mathbf{h}_j^*\|}\right\}.$$

(We used Schwarz's Inequality in the last step.)

By Equation 7.49*b*, we therefore have

$$\lambda_{\min}(\mathbf{B}_m^{-1}) \geq \sum_{n=1}^{k-1} \min_{j \in J_n} \|\mathbf{h}_j^*\|^2 \left\{\lambda_{\min}\left[\sum_{j \in J_n} \frac{\mathbf{h}_j^*\mathbf{h}_j^{*\prime}}{\|\mathbf{h}_j^*\|^2}\right] - 2qC_1 r(\mathscr{P})\right\}.$$

Since

$$\lambda_{\min}\left[\sum_{j \in J_n} \frac{\mathbf{h}_j\mathbf{h}_j^{*\prime}}{\|\mathbf{h}_j^*\|^2}\right] \leq \operatorname{tr}\left[\sum_{j \in J_n} \frac{\mathbf{h}_j^*\mathbf{h}_j^{*\prime}}{\|\mathbf{h}_j^*\|^2}\right] \leq q,$$

we see that

$$\lambda_{\min}(\mathbf{B}_m^{-1}) \geq \sum_{n=1}^{k} \min_{j \in J_n} \|\mathbf{h}_j^*\|^2 \left\{\lambda_{\min}\left[\sum_{j \in J_n} \frac{\mathbf{h}_j^*\mathbf{h}_j^{*\prime}}{\|\mathbf{h}_j^*\|^2}\right] - 2qC_1 r(\mathscr{P})\right\} - q \min_{j \in J_k} \|\mathbf{h}_j^*\|^2$$

if $m \in J_k$.

In much the same fashion,

$$\lambda_{\max}(\mathbf{B}_m^{-1}) \leq \sum_{n=1}^{k} \max_{j \in J_n} \|\mathbf{h}_j^*\|^2 \left\{\lambda_{\max}\left[\sum_{j \in J_n} \frac{\mathbf{h}_j^*\mathbf{h}_j^{*\prime}}{\|\mathbf{h}_j^*\|^2}\right] + q[C_1^2 r^2(\mathscr{P}) + C_1 r(\mathscr{P})]\right\} + \lambda_{\max}(\mathbf{R})$$

if $m \in J_k$.

Thus, if $m \in J_k$, we have

$$\kappa_m = \frac{\lambda_{\min}(\mathbf{B}_m^{-1})}{\lambda_{\max}(\mathbf{B}_m^{-1})} \geq$$

$$\frac{\sum_{n=1}^{k} \min_{j \in J_n} \|\mathbf{h}_j^*\|^2 \left\{\lambda_{\min}\left[\sum_{j \in J_n} \frac{\mathbf{h}_j^*\mathbf{h}_j^{*\prime}}{\|\mathbf{h}_j^*\|^2}\right] - 2qC_1 r(\mathscr{P})\right\} - q \min_{j \in J_k} \|\mathbf{h}_j^*\|^2}{\sum_{n=1}^{k} \max_{j \in J_n} \|\mathbf{h}_j^*\|^2 \left\{\lambda_{\max}\left[\sum_{j \in J_n} \frac{\mathbf{h}_j^*\mathbf{h}_j^{*\prime}}{\|\mathbf{h}_j^*\|^2}\right] + q[C_1^2 r^2(\mathscr{P}) + 2C_1 r(\mathscr{P})]\right\} + \lambda_{\max}(\mathbf{R})}.$$

By virtue of Assumptions F3, F4, and F5, the sums in the denominator and numerator approach $+\infty$, while the ratio of the second term in the numerator to the sum in the denominator approaches zero by Assumption F2. Using the discrete version of L'Hospital's rule, we find that

$$\liminf_{m \to \infty} \kappa_m \geq$$

$$\liminf_{k \to \infty} \frac{\min_{j \in J_k} \|\mathbf{h}_j^*\|^2 \left\{\lambda_{\min}\left[\sum_{j \in J_k} \mathbf{h}_j^*\mathbf{h}_j^{*\prime}/\|\mathbf{h}_j^*\|^2\right] - 2qC_1 r(\mathscr{P})\right\}}{\max_{j \in J_k} \|\mathbf{h}_j^*\|^2 \left\{\lambda_{\max}\left[\sum_{j \in J_k} \mathbf{h}_j^*\mathbf{h}_j^{*\prime}/\|\mathbf{h}_j^*\|^2\right] + q[C_1^2 r^2(\mathscr{P}) + 2C_1 r(\mathscr{P})]\right\}},$$

and by Assumptions F4, F5, and F6, we see that the last is greater than or equal to

$$\left(\frac{\tau^*}{\sigma^* K^q}\right)^2 - C_9 r(\mathscr{P}),$$

which proves Equation 7.67.

The proof of part d is in the same vein as b, and we leave the details to the reader. Q.E.D.

7.7 Limitations of the Recursive Method. Ill Conditioning

In the parlance of numerical analysis, a matrix $\mathbf{HH}'$ is said to be ill conditioned if

$$\lambda_{\max}(\mathbf{HH}')/\lambda_{\min}(\mathbf{HH}')$$

is large but finite. The column vectors of such matrices are "just barely" linearly independent, and, when one tries to compute the value of $\mathbf{x}$ that minimizes

$$\|\mathbf{z} - \mathbf{H}'\mathbf{x}\|^2$$

(that is, $\hat{\mathbf{x}} = (\mathbf{HH}')^{-1}\mathbf{Hz}$), one finds that the numerical solution is extremely sensitive to round-off errors (compare Householder, 1964, Chap. 5). The notion of ill conditioning extends naturally to the large-sample theory of recursive linear regression if we call a linear-regression function (actually a *sequence* of regression vectors) $\{\mathbf{h}_n\}$ *ill conditioned* whenever we have

$$\limsup_{n\to\infty} \frac{\lambda_{\max}\left(\sum_{j=1}^{n} \mathbf{h}_j\mathbf{h}_j'\right)}{\lambda_{\min}\left(\sum_{j=1}^{n} \mathbf{h}_j\mathbf{h}_j'\right)} = \infty. \tag{7.68}$$

This extension of the terminology is a reasonable one; for, if observations are made on a process of the form

$$Y_n = \mathbf{h}_n'\boldsymbol{\theta} + Z_n \qquad (n = 1, 2, \cdots),$$

and if we attempt to estimate $\boldsymbol{\theta}$ recursively (by means of Equation 7.9), it is necessary to compute $\mathbf{B}_n = (\sum_{j=1}^{n} \mathbf{h}_j\mathbf{h}_j')^{-1}$ at each step of the recursion. If $\{\mathbf{h}_n\}$ is ill conditioned, this computation becomes increasingly unstable with regard to round-off errors.

If $\{\mathbf{h}_n\}$ is ill conditioned, this does not preclude the possibility that $\lambda_{\max}(\mathbf{B}_n) \to 0$. Such situations are very perplexing from the practical point of view. On the one hand, theoretical considerations lead us to expect consistency from the recursively computed least-squares estimator (cf. Section 7.1). On the other hand, numerical considerations can easily cause the recursion to generate a nonsensical output.

The "classic" instance of such a situation arises in polynomial regression where $f_n(\boldsymbol{\theta}) = \sum_{j=0}^{p} \theta_j n^j$ and

$$\mathbf{h}_n = \begin{bmatrix} 1 \\ n \\ n^2 \\ \vdots \\ n^p \end{bmatrix}$$

In fact, polynomial regression is a particular instance of a more general class of ill-conditioned regression functions:

THEOREM 7.2

If $\sum_n \|\mathbf{h}_n\|^2 = \infty$ and $\lim_n \mathbf{h}_n/\|\mathbf{h}_n\| = \mathbf{h}$, then $\{\mathbf{h}_n\}$ is ill conditioned. (We defer the proof till the end of this section.) For instance, if

$$\mathbf{h}_n = \begin{bmatrix} 1 \\ n \end{bmatrix},$$

it is clear that

$$\frac{\mathbf{h}_n}{\|\mathbf{h}_n\|} \to \begin{bmatrix} 0 \\ 1 \end{bmatrix};$$

therefore, Theorem 7.2 applies. At the same time, we have

$$\mathbf{B}_n = \left(\sum_{k=1}^{n} \mathbf{h}_k \mathbf{h}_k'\right)^{-1} = \left[n \sum_{k=1}^{n} k^2 - \left(\sum_{k=1}^{n} k\right)^2\right]^{-1} \begin{bmatrix} \sum_{k=1}^{n} k^2 & \sum_{k=1}^{n} k \\ -\sum_{k=1}^{n} k & n \end{bmatrix}$$

The first factor on the right-hand side is less than a constant times n^{-4}; therefore, tr $\mathbf{B}_n = O(1/n) \to 0$. In cases such as these, we can only advise the practitioner to exercise extreme caution in designing his computational program.

In light of Lemma 7b, *ill-conditioned linear-regression functions must necessarily violate at least one of the hypotheses of Theorem 7.1.* If, in particular, the regression is ill conditioned owing to the fact that

$$\sum_n \|\mathbf{h}_n\|^2 = \infty, \quad \text{and} \quad \frac{\mathbf{h}_n}{\|\mathbf{h}_n\|} \to \mathbf{h},$$

it follows that

$$\lim_{n\to\infty} \lambda_{\min}\left(\sum_{j=n}^{n+k} \frac{\mathbf{h}_j\mathbf{h}_j'}{\|\mathbf{h}_j\|^2}\right) = k\lambda_{\min}(\mathbf{hh}') = 0$$

for any k, which means that *Assumptions C3, D3, and E3 of Chapter 6 are violated.* This of itself does not preclude consistency (for example, least-squares polynomial regression). However, the theorems of Chapters 6 and 7 don't apply. In particular, the "quick and dirty" recursion applied to polynomial regression cannot be shown to be consistent.

These observations apply even more strongly to the case of nonlinear regression. A nonlinear regression function exhibits the pathology of ill conditioning if Equation 7.68 holds when $\mathbf{h}_n$ is the gradient of the regression function evaluated at the true parameter value.

Proof of Theorem 7.2. Since det $\mathbf{A}$ is equal to the product of $\mathbf{A}$'s eigenvalues, it must be that

$$\det \mathbf{A} \geq \lambda_{\max}(\mathbf{A})[\lambda_{\min}(\mathbf{A})]^{p-1}$$

if $\mathbf{A}$ is $p \times p$ and nonnegative definite. On the other hand, we see that

$$\lambda_{\max}(\mathbf{A}) \geq \frac{1}{p}\operatorname{tr}(\mathbf{A}).$$

Combining these results, we find that

$$\frac{\lambda_{\max}^p(\mathbf{A})}{\lambda_{\min}^{p-1}(\mathbf{A})\lambda_{\max}(\mathbf{A})} = \left[\frac{\lambda_{\max}(\mathbf{A})}{\lambda_{\min}(\mathbf{A})}\right]^{p-1} \geq \frac{\left(\frac{1}{p}\operatorname{tr}(\mathbf{A})\right)^p}{\det(\mathbf{A})}. \tag{7.69a}$$

In the case at hand, we can write

$$\mathbf{h}_n/\|\mathbf{h}_n\| = \mathbf{h} + \mathbf{r}_n, \qquad \text{where } \|\mathbf{r}_n\| \to 0;$$

therefore,

$$\det\left(\sum_{j=1}^{n} \mathbf{h}_j\mathbf{h}_j'\right) = \det\left[\sum_{j=1}^{n} \|\mathbf{h}_j\|^2 (\mathbf{hh}' + \mathbf{r}_j\mathbf{h}' + \mathbf{hr}_j' + \mathbf{r}_j\mathbf{r}_j')\right].$$

Since

$$\operatorname{tr}\left(\sum_{j=1}^{n}\mathbf{h}_j\mathbf{h}_j'\right) = \sum_{j=1}^{n}\|\mathbf{h}_j\|^2,$$

it follows that

$$\frac{\det\left(\sum_{j=1}^{n} \mathbf{h}_j\mathbf{h}_j'\right)}{\left[\operatorname{tr}\left(\sum_{j=1}^{n}\mathbf{h}_j\mathbf{h}_j'\right)\right]^p} = \det\left[\mathbf{hh}' + \frac{\sum_{j=1}^{n}\|\mathbf{h}_j\|^2(\mathbf{r}_j\mathbf{h}' + \mathbf{hr}_j' + \mathbf{r}_j\mathbf{r}_j')}{\sum_{j=1}^{n}\|\mathbf{h}_j\|^2}\right]$$

$$= \det[\mathbf{hh}' + \mathbf{R}_n]. \tag{7.69b}$$

But

$$\|\mathbf{R}_n\| \le \frac{\sum_{j=1}^{n} \|\mathbf{h}_j\|^2[2\|\mathbf{h}\|\,\|\mathbf{r}_j\| + \|\mathbf{r}_j\|^2]}{\sum_{j=1}^{n} \|\mathbf{h}_j\|^2}.$$

Since

$$\sum_{j=1}^{n} \|\mathbf{h}_j\|^2 \to \infty,$$

the discrete version of L'Hospital's rule applies. Since $\|\mathbf{r}_n\| \to 0$, it follows that $\|\mathbf{R}_n\| \to 0$, from which it follows that

$$\det[\mathbf{h}\mathbf{h}' + \mathbf{R}_n] \to \det[\mathbf{h}\mathbf{h}'] = 0. \qquad \textbf{(7.69}\textbf{\textit{c}}\textbf{)}$$

We combine Equations 7.69*a*, *b*, and *c*, and the theorem is established. Q.E.D.

7.8 Response Surfaces

Until now, we have motivated recursive-estimation procedures by considering regression problems in the setting of time-series analysis. It is in these applications that the demands of "on-line" computation make recursive-estimation techniques particularly attractive. In such cases, the regression function is typically of the form

$$F_n(\boldsymbol{\theta}) = F(t_n; \boldsymbol{\theta}) \qquad (n = 1, 2, \cdots),$$

where, for each $\boldsymbol{\theta}$, $F(\cdot\,; \boldsymbol{\theta})$ is a continuous function of time, and

$$t_1 < t_2 < \cdots < t_n \to \infty \qquad (7.70)$$

are the sampling instants. The large-sample properties of recursive-estimation sequences are determined by the analytic properties of $F(\cdot\,; \boldsymbol{\theta})$ for large values of t.

However, the scope of regression analysis also embraces experimental situations where the regression function is of the form $F(\mathbf{t}; \boldsymbol{\theta})$, $\mathbf{t}$ now denoting a (possibly abstract) variable that the experimenter can choose more or less at will (with replication if so desired) from a certain set of values. In particular, the constraint of 7.70 is not present. In fact, the values of the independent variable $\mathbf{t}$ are usually chosen from a set that is bounded (in an appropriate metric) or compact (in an appropriate topology). For example, $F(\mathbf{t}; \boldsymbol{\theta})$ might be the mean yield of a chemical process when the control variables (temperature, pressure, input quantities, and so on) are represented by the vector $\mathbf{t}$ and the external variables (not under the control of the experimenter, and, indeed, generally unknown) are denoted by $\boldsymbol{\theta}$.

In such cases (where $\mathbf{t}$ is a finite dimensional vector), the regression function describes a surface that is indexed by $\boldsymbol{\theta}$ as $\mathbf{t}$ varies over its domain. This surface is called a *response surface*, and if $\boldsymbol{\theta}$ is not known, the job of fitting the proper response surface to data $Y_1, Y_2, Y_3, \cdots$ (which are noisy observations taken at settings $\mathbf{t}_1, \mathbf{t}_2, \mathbf{t}_3, \cdots$ of the independent control variable) is equivalent to choosing the "correct" value of $\boldsymbol{\theta}$ on the basis of the noisy observations

$$Y_n = F(\mathbf{t}_n; \boldsymbol{\theta}) + W_n \qquad (n = 1, 2, \cdots).$$

In most cases, the experimenter wishes to estimate $\boldsymbol{\theta}$ once and for all after all the data have been collected. One could apply the recursive method, but its chief selling point, the availability of a *running* estimate for $\boldsymbol{\theta}$, is of no great value. However, when questions of *sequential* experimentation arise, this feature regains its allure.

For example, suppose the experimenter wishes to determine the correct response surface, and suppose he can make observations using either of two different experimental procedures. Under the first procedure, his observations take the form

$$V_n = F(\mathbf{t}_n; \boldsymbol{\theta}) + v_n \qquad (n = 1, 2, \cdots),$$

the v_n being independent, zero-mean measurement errors with variance σ_v^2. The second procedure generates observations of the form

$$U_n = F(\mathbf{t}_n; \boldsymbol{\theta}) + u_n \qquad (n = 1, 2, \cdots),$$

the u_n being independent, zero-mean errors with variance σ_u^2. If σ_u^2 and σ_v^2 were known, and if data were costly to obtain, it is clear that a sophisticated experimentalist would choose the observation procedure with the smaller variance and use it exclusively. However, if the variances are not known *a priori*, a sensible thing to do is to allocate some experimental effort to estimate σ_v^2 and σ_u^2 and then sample from the population with the lower variance estimate. Or, one could proceed sequentially, sampling from each population according to the outcome of a chance device, whose probability law increasingly favors the population with the lower variance estimate. Such a procedure requires a running estimate for $\boldsymbol{\theta}$ in order that the variance estimates be computable after each new observation. [Actually, the sequential design of experiments demands a far more sophisticated approach, but the present oversimplified procedure suffices to motivate the application of recursive methods to the fitting of response surfaces. The interested reader is advised to refer to Chernoff's paper (1959) for a proper introduction to sequential experimentation.]

From the theoretical point of view, the most appealing feature of the recursive method, applied to the determination of response surfaces, is the wide class of regressions (apparently much larger than those in time-series applications) that satisfy the hypotheses of Theorem 7.1. The following theorem demonstrates the great simplifications that obtain when the independent variable $\mathbf{t}$ is constrained to a compact set.

THEOREM 7.3

Let $\mathscr{T}$ be a compact set, let $\mathscr{P}$ be a convex, compact subset of p-dimensional Euclidean space, and suppose that $F(\cdot\,;\,\cdot)$ is a real-valued function defined over $\mathscr{T} \otimes \mathscr{P}$, having the following properties:

G1. $\partial^2 F/\partial\theta_i\partial\theta_j$ exists and is continuous over $\mathscr{T} \otimes \mathscr{P}$, and

G2. $\|\dot{\mathbf{F}}\|$ is continuous and positive on $\mathscr{T} \otimes \mathscr{P}$, where $\dot{\mathbf{F}}$ is the column vector whose components are

$$\frac{\partial F}{\partial \theta_i} \qquad (i = 1, 2, \cdots, p).$$

Let $\mathbf{t}_1, \mathbf{t}_2, \cdots$ be a sequence of points from $\mathscr{T}$ and let

$$F_n(\mathbf{x}) = F(\mathbf{t}_n; \mathbf{x}).$$

G3. If there exists a sequence of integers

$$1 = \nu_1 < \nu_2 < \cdots$$

with

$$p \le \nu_{k+1} - \nu_k = p_k \le q < \infty$$

such that

$$\liminf_{k\to\infty} \det\left(\sum_{j \in J_k} \dot{\mathbf{F}}_j(\mathbf{x})\dot{\mathbf{F}}_j'(\mathbf{x})\right) = D^2 > 0$$

for some $\mathbf{x} \in \mathscr{P}$, then Conclusion a of Theorem 7.1 holds.

Proof. $(\partial^2 F/\partial\theta_i\partial\theta_j)$ is continuous on $\mathscr{T} \otimes \mathscr{P}$, which is compact; therefore, $|(\partial^2 F/\partial\theta_i\partial\theta_j)|$ is uniformly bounded. Therefore $|(\partial^2 F_n/\partial\theta_i\partial\theta_j)|$ is uniformly bounded in i, j, and n. The continuity and positivity of $\|\dot{\mathbf{F}}\|$ over $\mathscr{T} \otimes \mathscr{P}$ implies the existence of positive K_1 and K_2, such that

$$K_1 \le \|\dot{\mathbf{F}}_n(\mathbf{x})\| \le K_2$$

for all n and all $\mathbf{x} \in \mathscr{P}$. These facts establish Assumptions F1 through F4.

If $\mathbf{B}$ is $p \times p$ and nonnegative definite, we find that

$$\det(\mathbf{B}) \le \lambda_{\min}(\mathbf{B})\lambda_{\max}^{p-1}(\mathbf{B}) \le \lambda_{\min}(\mathbf{B})[\operatorname{tr}(\mathbf{B})]^{p-1}.$$

Since

$$\lambda_{\min}\left(\sum_{j\in J_k}\frac{\dot{\mathbf{F}}_j\dot{\mathbf{F}}_j'}{\|\dot{\mathbf{F}}_j\|^2}\right) \geq \frac{1}{K_2^2}\lambda_{\min}\left(\sum_{j\in J_k}\dot{\mathbf{F}}_j\dot{\mathbf{F}}_j'\right),$$

and since

$$\operatorname{tr}\left(\sum_{j\in J_k}\dot{\mathbf{F}}_j\dot{\mathbf{F}}_j'\right) = \sum_{j\in J_k}\|\dot{\mathbf{F}}_j\|^2 \leq qK_2^2,$$

we conclude that Assumption F5 holds if Assumption G3 holds.

Comment: The set $\mathscr{T}$ may be abstract, compact with respect to an arbitrary topology, provided that $F(\cdot\,;\,\cdot)$ is continuous on $\mathscr{T}\otimes\mathscr{P}$ in the induced product topology. However, in most (but not all) applications, $\mathscr{T}$ will be a closed bounded subset of some finite dimensional Euclidean space.

We close this chapter by exhibiting examples of regression functions of the form $F_n(\boldsymbol{\theta}) = F(t_n;\boldsymbol{\theta})$ which violate the conditions that justify the recursive method if $t_n \to \infty$, but which satisfy the conditions of Theorem 7.3 if the t_n are chosen appropriately from a finite interval.

Example 7.1. It has been shown in Chapter 2 that the regression $F(t_n;\theta) = \cos\theta t_n$ or $\sin\theta t_n$ violates the conditions of Theorem 2.1 if $t_n = n$ (or $n\tau$). However, if θ is known to lie in an interval $0 < \alpha_2 \leq \theta \leq \beta_2$ and $\{t_n\}$ is a suitably chosen sequence, the difficulty disappears. In fact, to make the problem more interesting, consider

$$F(t_n;\boldsymbol{\theta}) = \theta_1\sin\theta_2 t_n,$$

where

$$\mathscr{P} = [\alpha_1,\beta_1]\otimes[\alpha_2,\beta_2] \qquad (\alpha_1,\beta_1 > 0),$$

$$0 < T_1 = \inf_n t_n < \sup_n t_n = T_2 < \pi/2\beta_2,$$

and

$$\inf_n (t_{2n} - t_{2n-1}) = T_3 > 0.$$

The function

$$F(t,\boldsymbol{\theta}) = \theta_1\sin\theta_2 t$$

satisfies Assumptions G1 and G2 over $\mathscr{T}\otimes\mathscr{P}$ if we take $\mathscr{T} = [T_1, T_2]$. On the other hand, a little algebra shows that

$$\det[\dot{\mathbf{F}}_{2n+1}(\boldsymbol{\theta})\dot{\mathbf{F}}'_{2n+1}(\boldsymbol{\theta}) + \dot{\mathbf{F}}_{2n}(\boldsymbol{\theta})\dot{\mathbf{F}}'_{2n}(\boldsymbol{\theta})]$$

$$= \left[\frac{\theta_1^2\theta_2^2}{4}(t_{2n+1}^2 - t_{2n}^2)\right]\left[\frac{\sin\theta_2(t_{2n+1}-t_{2n})}{\theta_2(t_{2n+1}-t_{2n})} - \frac{\sin\theta_2(t_{2n+1}+t_{2n})}{\theta_2(t_{2n+1}+t_{2n})}\right]^2.$$

The first factor is larger than $\frac{1}{2}(\alpha_1\alpha_2)^2T_1T_3$. If we let

$$\mu_n = \theta_2(t_{2n+1} - t_{2n}),$$

and

$$\omega_n = \theta_2(t_{2n+1} + t_{2n}),$$

we see that

$$0 < \alpha_2 T_3 \le \mu_n \le \omega_n - 2\alpha_2 T_1 < \omega_n \le 2T_2\beta_2 < \pi.$$

Since $\sin \xi/\xi$ has a negative derivative which is bounded away from zero in the interval $[\alpha_2 T_3, 2T_2\beta_2]$, we find that

$$\inf_n \left(\frac{\sin \mu_n}{\mu_n} - \frac{\sin \omega_n}{\omega_n}\right)^2 > 0.$$

This establishes Assumption G3 if we take $\nu_k = 2k - 1$ $(k = 1, 2, \cdots)$.

Example 7.2. It was also shown in Chapter 2 that the exponential regression $e^{\theta t_m}$ violates the conditions of Theorem 2.1 in an essential way if $t_n = n\tau$. But consider the more general regression

$$F(t_n; \theta) = e^{\theta_1 t_n} + e^{\theta_2 t_n},$$

and suppose it is known that $-\infty < \alpha_1 \le \theta_1 \le \beta_1 < \alpha_2 \le \theta_2 \le \beta_2 < \infty$. If the sampling instants are chosen so that

$$0 < T_1 = \inf_n t_n < \sup_n t_n = T_2 < \infty,$$

and

$$\inf_n (t_{2n+1} - t_{2n}) = T_3 > 0,$$

then Assumptions G1 and G2 hold for the function

$$F(t; \boldsymbol{\theta}) = e^{\theta_1 t} + e^{\theta_2 t}$$

defined over

$$\mathscr{T} \otimes \mathscr{P} = [T_1, T_2] \otimes [\alpha_1, \beta_1] \otimes [\alpha_2, \beta_2].$$

Furthermore,

$$\begin{aligned}
&\det(\dot{\mathbf{F}}_{2n+1}\dot{\mathbf{F}}'_{2n+1} + \dot{\mathbf{F}}_{2n}\dot{\mathbf{F}}'_{2n})\\
&\quad = (t_{2n+1}t_{2n})^2[\exp 2(\theta_2 t_n + \theta_1 t_{2n+1})][1 - \exp(\theta_2 - \theta_1)(t_{2n+1} - t_{2n})]^2\\
&\quad \ge T_1^4[\exp 2(\alpha_2 t_n + \alpha_1 t_{2n+1})][1 - \exp(\alpha_2 - \beta_1)T_3]^2.
\end{aligned}$$

Since $\{t_n\}$ is a bounded sequence, the second factor is bounded away from zero; therefore, Assumption G3 holds with $\nu_k = 2k - 1$ $(k = 1, 2, \cdots)$.

Another generalized version of the exponential regression is given by

$$F(t_n; \boldsymbol{\theta}) = \theta_1 e^{\theta_2 t_n}.$$

If we assume that

$$\boldsymbol{\theta} \in \mathscr{P} = [\alpha_1, \beta_1] \otimes [\alpha_2, \beta_2]$$

where the α's and β's can be positive or negative, and if the sampling instants are chosen so that

$$T_1 = \inf_n t_n < \sup_n t_n = T_2, \qquad \text{and} \qquad 0 < T_3 = \inf_n (t_{2n+1} - t_{2n}),$$

then

$$F(t; \boldsymbol{\theta}) = \theta_1 e^{\theta_2 t}$$

satisfies Assumptions G1 and G2 on $[T_1, T_2] \otimes \mathscr{P}$. Moreover, we see that

$$\det (\dot{\mathbf{F}}_{2n+1}\dot{\mathbf{F}}'_{2n+1} + \dot{\mathbf{F}}_{2n}\dot{\mathbf{F}}'_{2n}) = \theta_1^2 \Big[\exp [2\theta_2(t_{2n+1} - t_{2n})]\Big](t_{2n+1} - t_{2n})^2$$

$$\geq \alpha_1^2 T_3^2 \exp 2\alpha_2(t_{2n+1} - t_{2n}),$$

and the last is bounded away from zero since the t_n are bounded. Thus, Assumption G3 holds for this regression as well.

Example 7.3. In Section 7.7, we showed that polynomial regressions

$$F(t_n; \boldsymbol{\theta}) = \sum_{j=0}^{p} \theta_j t_n^j$$

fail to satisfy Assumptions C3, D3, and E3 of Chapter 6 if $t_n \to \infty$. However, if the sampling instants are suitably chosen from a compact set, this difficulty also evaporates. To illustrate this, consider the case of a first-degree polynomial

$$F(t; \boldsymbol{\theta}) = \theta_0 + \theta_1 t$$

sampled at times $\{t_n\}$ in the interval $[T_1, T_2]$ and having the property that

$$\inf_n (t_{2n+1} - t_{2n}) = T_3 > 0.$$

Letting

$$F_n(\boldsymbol{\theta}) = F(t_n; \boldsymbol{\theta}),$$

we find, as usual, that

$$\det (\dot{\mathbf{F}}_{2n+1}\dot{\mathbf{F}}'_{2n+1} + \dot{\mathbf{F}}_{2n}\dot{\mathbf{F}}'_{2n}) = (t_{2n+1} - t_{2n})^2 \geq T_3.$$

Assumptions G1 and G2 are satisfied over any compact $\boldsymbol{\theta}$-set; therefore, the conclusions of Theorem 7.3 hold. In particular, notice that the problem is no longer ill conditioned.

In closing, we point out that all three examples require that samples be made over a bounded interval, in such a way that

$$t_{2n+1} - t_{2n} \geq T_3 > 0.$$

One such scheme (defined over the interval [0, 1], with $T_3 = \frac{1}{2}$) chooses $t_1 = \frac{1}{2}$ and

$$\left.\begin{aligned} t_{2^k+2(j-1)} &= \frac{2j-1}{2^{k+1}} \\ t_{2^k+2j-1} &= \frac{1}{2} + \frac{2j-1}{2^{k+1}} \end{aligned}\right\} \qquad (j = 1, 2, \cdots, 2^{k-1};\ k = 1, 2, \cdots).$$

8. Applications

Before we can apply the recursive methods described in Chapters 6 and 7 to particular regression problems, several decisions must be made. Should the data be processed one observation at a time or in batches? In the latter case, what should the batch sizes be? Should the recursion be truncated? Which type of gains should be used, deterministic or adaptive?

At this writing, definitive answers to these questions are not available. It seems clear, however, that each issue should be weighed in the context of the consistency and computability of the resulting estimator-recursion.

Consistency (in either the mean-square or almost-sure sense) is the most important consideration. A procedure for which consistency cannot be established (or conjectured with high certainty) should be held in less esteem than one to which the theorems of Chapters 6 and 7 apply. This comment is particularly relevant to the decision concerning truncation. For example, suppose that a particular gain sequence is being contemplated and that one or more of Assumptions C1 through C6′ of Theorem 6.2 are violated. However, suppose it is known *a priori* that the true parameter value $\boldsymbol{\theta}$ lies inside a given sphere $\mathscr{P}$, which is sufficiently small so that Assumptions E1 through E6′ are satisfied. (This situation often arises in practice.) If an untruncated procedure were used, convergence could not be ensured. However, a batch-processing recursion, truncated over $\mathscr{P}$ with batch sizes chosen so that Assumptions

E3 through E5 hold, does converge to $\boldsymbol{\theta}$ in the mean square (by Theorem 6.4). If data-processing considerations make batch processing unfeasible, the single-observation recursion, truncated over $\mathscr{P}$, appears to be the natural alternative in such a case. Although the theorem concerning convergence of truncated single-observation recursions is conjectural, we are confident enough in its validity to feel safe in recommending it under the above-mentioned circumstances.

In some applications, single-observation recursions may be dictated by cost considerations. If observations are very expensive and we are estimating $\boldsymbol{\theta}$ as part of a sequential hypothesis-testing procedure, the single-observation recursion is the natural one to choose. In other applications, the data may be collected in batches (for example, multiple sensors reporting simultaneously) and should be so processed.

If one is confronted with a situation where a free choice exists, we tend to favor batch processing. From the theoretical point of view, it would appear that once the gains are decided on, the batch sizes should be chosen to make Assumptions E3 through E5 hold. In all likelihood, though, the choice of batch size is of no practical consequence and can be chosen purely for convenience (ideally, though, as *large* as computationally convenient).

The considerations governing the choice of gain sequence are clear cut in the case of linear regression but not so well defined in the nonlinear case. In the linear case, the gains of the type 7.9 or 7.39, depending on whether the single-observation or batch-processing recursion is used, are preferred unless the rate of data acquisition is so high that the data-processing facility is swamped. In this case, the "quick and dirty" gains of Equations 7.18 and 7.40 yield estimator sequences which can be computed more quickly and thus can better keep pace with the data. One sacrifices statistical efficiency by doing so, though, and in order to guarantee consistency, two additional conditions (Equations 7.19*a*, *b*) must be verified.

For nonlinear regression, the decision is more delicate. Referring to Theorem 7.1, we see that the "quick and dirty" gains (7.48*a*, *c*) yield convergence under a set of conditions which are weaker than those required by the "linearized least-squares" gains (7.48*b*, *d*). If the additional Assumption F6 can be verified, the choice between gains of the type 7.48*a*, *c* and 7.48*b*, *d* involves the weighing of efficiency versus computability. (The "linearized least-squares" gain probably yields a more efficient recursion, since it entails more computation.) If, on the other hand, the extra condition cannot be verified, the "linearized least-squares" gains may not yield a convergent estimator. Thus, the use of the "quick and dirty" gain is the conservative course of action.

Alternatively, one might use the "quick and dirty" gain initially to get things started and then switch to the other type of gain, under the supposition that the linearized version of the problem is, by then, an adequate approximation. This approach can be investigated analytically in the spirit of the present work, but we will not pursue it further.

If the "linearized least-squares" gains of Equations 7.48*b*, *d* are to be used, the results of the scalar-parameter case presented in Theorem 4.2 for Gains 2 and 3 show that we cannot state *a priori* that the adaptive version will be more efficient than the deterministic version (as one might expect). At this time, we can offer little in the way of guidelines for choosing between adaptive and deterministic linearized least-squares gains. However, adaptive gains must be computed after each cycle of the recursion and so, if pressed for time, we may be compelled to resort to the quasi-adaptive or deterministic versions. On the other hand, if "quick and dirty" gains are being used because of time considerations, the sensible thing to do is to use the deterministic versions. These can be stored in memory and need not be computed in real-time. If the "quick and dirty" gains are being used because Assumption F6 of Theorem 7.1 cannot be established, the adaptive version might conceivably speed up the convergence rate somewhat.

We will now display some examples and show, in each case, how to go about verifying the conditions which will guarantee consistency of the recursive-estimation procedure used.

8.1 Vector Observations and Time-Homogeneous Regression

Example 8.1. Suppose the observations are an r-dimensional vector stochastic process of the form

$$\mathbf{Y}_k = \mathbf{f}(\boldsymbol{\theta}) + \mathbf{Z}_k \qquad (k = 1, 2, \cdots),$$

where the components of the residual vectors have uniformly bounded variances. An estimate for the unknown p-vector ($p \leq r$) $\boldsymbol{\theta}$ is sought. The "classical" approach would involve estimating the mean-value vector $\mathbf{f}(\boldsymbol{\theta})$ by

$$\hat{\mathbf{f}}_n = \frac{1}{n} \sum_{j=1}^{n} \mathbf{Y}_j$$

and solving (by least squares perhaps) for the value of $\hat{\boldsymbol{\theta}}_n$ that "comes closest" (in some sense) to making the equations

$$\mathbf{f}(\hat{\boldsymbol{\theta}}_n) = \hat{\mathbf{f}}_n$$

work.

By contrast, the recursive approach estimates $\boldsymbol{\theta}$ directly:

$$\mathbf{s}_{k+1} = \mathbf{s}_k + \mathbf{A}_k[\mathbf{Y}_k - \mathbf{f}(\mathbf{s}_k)].$$

The batch-processing recursion is the natural one to use when the "observations" are vectors to begin with.

On the other hand, if the components of $\mathbf{Y}_k$ are observed one at a time, we could write

$$\mathbf{Y}_k = \begin{bmatrix} Y_{(k-1)r+1} \\ \vdots \\ Y_{kr} \end{bmatrix}, \qquad \mathbf{Z}_k = \begin{bmatrix} Z_{(k-1)r+1} \\ \vdots \\ Z_{kr} \end{bmatrix}$$

and

$$\mathbf{f}(\boldsymbol{\theta}) = \begin{bmatrix} f_1(\boldsymbol{\theta}) \\ \vdots \\ f_r(\boldsymbol{\theta}) \end{bmatrix},$$

so that

$$Y_n = F_n(\boldsymbol{\theta}) + Z_n \qquad (n = 1, 2, \cdots),$$

where

$$F_{kr+i}(\boldsymbol{\theta}) = f_i(\boldsymbol{\theta}) \qquad (i = 1, 2, \cdots, r; \quad k = 0, 1, \cdots). \tag{8.1}$$

In this case, we could justifiably consider the single-observation recursion. However, for the purposes of this example, we confine our attention to the batch-processing recursion.

We will assume the following:

$\boldsymbol{\theta}$ is known to lie inside a prescribed p-dimensional sphere $\mathscr{P}$. (8.2)

The components of each of the vector-valued functions

$\dot{\mathbf{f}}_i(\cdot) = \operatorname{grad} f_i(\cdot) \qquad (i = 1, 2, \cdots, r)$
are continuously differentiable over $\mathscr{P}$. (8.3)

For each $\mathbf{x} \in \mathscr{P}$, the set of vectors $\dot{\mathbf{f}}_1(\mathbf{x})$ $\dot{\mathbf{f}}_2(\mathbf{x}), \cdots, \dot{\mathbf{f}}_r(\mathbf{x})$
has rank p and all have positive lengths. (8.4)

We also assume that either

$\{Z_n\}$ is an independent (scalar) process with mean zero (8.5)

or

$$\mathscr{E}Z_n^2 = O(n^{-\delta}) \qquad \text{for some } \delta > 0. \tag{8.6}$$

We will consider the (truncated, batch-processed) recursion

$$\begin{aligned} \mathbf{s}_{k+1} &= [\mathbf{s}_k + \mathbf{A}_k(\mathbf{Y}_k - \mathbf{f}(\mathbf{s}_k))]_{\mathscr{P}}, \\ \mathbf{s}_1 &= \boldsymbol{\theta}_0 \in \mathscr{P}, \end{aligned} \tag{8.7}$$

where $\mathbf{A}_k$ can be any of the following $p \times r$ matrices:

$$\mathbf{A}_k = \frac{1}{k}\left(\sum_{j=1}^{r} \|\dot{\mathbf{f}}_j(\boldsymbol{\theta}_0)\|^2\right)^{-1}(\dot{\mathbf{f}}_1(\boldsymbol{\theta}_0), \cdots, \dot{\mathbf{f}}_r(\boldsymbol{\theta}_0))$$

(deterministic, "quick and dirty"), (8.8*a*)

$$\mathbf{A}_k = \frac{1}{k}\left(\sum_{j=1}^{r} \|\dot{\mathbf{f}}_j(\mathbf{s}_k)\|^2\right)^{-1}(\dot{\mathbf{f}}_1(\mathbf{s}_k), \cdots, \dot{\mathbf{f}}_r(\mathbf{s}_k))$$

(adaptive, "quick and dirty"), (8.8*b*)

$$\mathbf{A}_k = \frac{1}{k}\left(\sum_{j=1}^{r} \dot{\mathbf{f}}_j(\boldsymbol{\theta}_0)\dot{\mathbf{f}}_j'(\boldsymbol{\theta}_0)\right)^{-1}(\dot{\mathbf{f}}_1(\boldsymbol{\theta}_0), \cdots, \dot{\mathbf{f}}_r(\boldsymbol{\theta}_0))$$

(deterministic, "linearized least-squares"), (8.8*c*)

$$\mathbf{A}_k = \frac{1}{k}\left(\sum_{j=1}^{r} \dot{\mathbf{f}}_j(\mathbf{s}_k)\dot{\mathbf{f}}_j'(\mathbf{s}_k)\right)^{-1}(\dot{\mathbf{f}}_1(\mathbf{s}_k), \cdots, \dot{\mathbf{f}}_r(\mathbf{s}_k))$$

(adaptive, "linearized least-squares"). (8.8*d*)

We will verify, in detail, that Equation 8.8*a* furnishes a mean-square convergent-estimator sequence and will sketch the arguments which are relevant to the corresponding proofs for Equations 8.8*b*, *c*, and *d*.

Let

$$\nu_k = (k-1)r + 1, \qquad \text{and} \qquad J_k = \{\nu_k, \nu_k + 1, \cdots, \nu_{k+1} - 1\},$$

so that the number of indices in J_k is

$$p_k = r \qquad (k = 1, 2, \cdots),$$

and let the column vectors of $\mathbf{A}_k$ be denoted by

$$\mathbf{a}_n \qquad (\nu_k \le n < \nu_{k+1}),$$

where

$$\mathbf{a}_n = \frac{1}{k}\left(\sum_{j=1}^{r} \|\mathbf{h}_j\|^2\right)^{-1}\mathbf{h}_n,$$

$$= \left(\sum_{j=1}^{\nu_{k+1}-1} \|\mathbf{h}_j\|^2\right)^{-1}\mathbf{h}_n, \qquad (\nu_k \le n < \nu_{k+1}),$$

and

$$\mathbf{h}_j = \dot{\mathbf{F}}_j(\boldsymbol{\theta}_0) \qquad (j = 1, 2, \cdots).$$

[$F_j(\cdot)$ is defined by Equation 8.1.] Under Equation 8.3, the matrices $\mathbf{G}_n(\mathbf{x}_1, \cdots, \mathbf{x}_p)$ whose column vectors are

$$\begin{bmatrix} \partial^2 F_n(\mathbf{y})/\partial y_1 \partial y_i \\ \vdots \\ \partial^2 F_n(\mathbf{y})/\partial y_p \partial y_i \end{bmatrix}_{\mathbf{y}=\mathbf{x}_i} \qquad (i = 1, 2, \cdots, p), \tag{8.9}$$

are uniformly bounded in norm as $(\mathbf{x}_1, \mathbf{x}_2, \cdots, \mathbf{x}_p)$ varies over the compact set $\mathscr{P}^{(p)}$. By Equation 8.4,

$$\|\mathbf{h}_n\| = \|\dot{\mathbf{F}}_n(\boldsymbol{\theta}_0)\| \geq \min_{i=1,2,\cdots,r} \|\dot{\mathbf{f}}_i(\boldsymbol{\theta}_0)\| > 0;$$

therefore, Assumptions F1 and F3 of Theorem 7.1 are satisfied. On the other hand,

$$\|\mathbf{h}_n\| \leq \max_{i=1,\cdots,r} \|\dot{\mathbf{f}}_i(\boldsymbol{\theta}_0)\| < \infty;$$

therefore, Assumptions F2 and F4 hold. Since

$$\lambda_{\min}\left(\sum_{j\in J_k} \frac{\mathbf{h}_j\mathbf{h}_j'}{\|\mathbf{h}_j\|^2}\right) \geq \max_{i=1,2,\cdots,r} \|\dot{\mathbf{f}}_i(\boldsymbol{\theta}_0)\|^{-2}\lambda_{\min}\left(\sum_{j=1}^{r} \dot{\mathbf{f}}_j(\boldsymbol{\theta}_0)\dot{\mathbf{f}}_j'(\boldsymbol{\theta}_0)\right),$$

Equation 8.4 implies Assumption F5. Since the gains are given by Equation 7.48*c*, it therefore follows from Theorem 6.4 (via Theorem 7.1*a*), that $\mathbf{s}_n$ converges to $\boldsymbol{\theta}$ if $\mathscr{P}$ is small enough and Equation 8.5 holds. If, instead of Equation 8.5, we assume Equation 8.6, Assumptions E1 through E6 are established as follows:

Let

$$\left.\begin{aligned}
&\mathbf{Y}_k^* = k^\delta \mathbf{Y}_k, \qquad \mathbf{F}_k^*(\mathbf{x}) = k^\delta \mathbf{f}(\mathbf{x}), \qquad \mathbf{Z}_k^* = k^\delta \mathbf{Z}_k,\\
&\mathbf{h}_n^* = k^\delta \mathbf{h}_n = k^\delta \dot{\mathbf{F}}_n(\boldsymbol{\theta}_0) \qquad (\nu_k \leq n < \nu_{k+1}),\\
&\mathbf{A}_k^* = \text{the } p \times r \text{ matrix whose column vectors are}\\
&\mathbf{a}_n^* = \left(\sum_{j=1}^{\nu_{k+1}-1} \|\mathbf{h}_j^*\|^2\right)^{-1} \mathbf{h}_n^* \qquad (\nu_k \leq n < \nu_{k+1}),
\end{aligned}\right\} \tag{8.10}$$

and consider the recursion

$$\mathbf{s}_{k+1}^* = [\mathbf{s}_k^* + \mathbf{A}_k^*(\mathbf{Y}_k^* - \mathbf{F}_k^*(\mathbf{s}_k^*))]_{\mathscr{P}}, \qquad \mathbf{s}_1^* = \boldsymbol{\theta}_0. \tag{8.11}$$

If we can show that $\mathbf{s}_k^* \to \boldsymbol{\theta}$ in quadratic mean, we will be done. This is so because the Corollary to Theorem 6.4 will then guarantee the mean-square convergence of

$$\hat{\mathbf{s}}_{k+1} = [\hat{\mathbf{s}}_k + \varphi_k\mathbf{A}_k^*(\mathbf{Y}_k^* - \mathbf{F}_k^*(\hat{\mathbf{s}}_k))]_{\mathscr{P}}, \qquad \hat{\mathbf{s}}_1 = \boldsymbol{\theta}_0, \tag{8.12}$$

with

$$\varphi_k = \sum_{j=1}^{k} \left(\frac{j^{2\delta}}{k^{1+2\delta}}\right).$$

Since

$$\varphi_k\mathbf{A}_k^*(\mathbf{Y}_k^* - \mathbf{F}_k^*(\hat{\mathbf{s}}_k)) = \mathbf{A}_k(\mathbf{Y}_k - \mathbf{F}_k(\hat{\mathbf{s}}_k))$$

for every k, the recursions of Equations 8.12 and 8.7 (hence $\mathbf{s}_k$ and $\hat{\mathbf{s}}_k$) are identical, which immediately establishes the mean-square convergence of $\mathbf{s}_k$ under Equation 8.6.

To establish Equation 8.11, notice that

$$\frac{n}{r} < k \leq \frac{n-1}{r} + 1$$

if

$$(k-1)r + 1 = \nu_k \leq n \leq \nu_{k+1} - 1 = kr.$$

Thus by Equation 8.10,

$$\left(\frac{n}{r}\right)^{\delta} \|\mathbf{h}_n\| \leq \|\mathbf{h}_n^*\| \leq \left(\frac{n}{r} + 1\right)^{\delta} \|\mathbf{h}_n\|, \tag{8.13}$$

and

$$\|\mathbf{G}_n^*\| \leq \left(\frac{n}{r} + 1\right)^{\delta} \|\mathbf{G}_n\|, \tag{8.14}$$

where $\mathbf{G}_n^*(\mathbf{x}_1, \cdots, \mathbf{x}_p)$ is the matrix whose columns are given by Equation 8.9, with F_n replaced by $k^{\delta}F_n$ ($\nu_k \leq n < \nu_{k+1}$). Thus, $\mathbf{F}_k^*$ satisfies Assumption F1 of Theorem 7.1, since

$$\frac{\|\mathbf{G}_n^*\|}{\|\mathbf{h}_n^*\|} \leq \left(\frac{n+r}{n}\right)^{\delta} \frac{\|\mathbf{G}_n\|}{\|\mathbf{h}_n\|},$$

the right-hand side being uniformly bounded by virtue of an earlier argument. Under Equations 8.3 and 8.4,

$$0 < \min_{i=1,\cdots,r} \|\dot{\mathbf{f}}_i(\boldsymbol{\theta}_0)\| \leq \|\mathbf{h}_j\| \leq \max_{i=1,\cdots,r} \|\dot{\mathbf{f}}_i(\boldsymbol{\theta}_0)\| < \infty;$$

therefore, by Equation 8.13,

$$K_1 n^{\delta} \leq \|\mathbf{h}_n^*\| \leq K_2 n^{\delta},$$

which establishes Assumption F2′. Since

$$\sum_{j \in J_k} \frac{\mathbf{h}_j^* \mathbf{h}_j^{*\prime}}{\|\mathbf{h}_j^*\|^2} = \sum_{j \in J_k} \frac{\mathbf{h}_j \mathbf{h}_j^{\prime}}{\|\mathbf{h}_j\|^2},$$

Assumption F5 holds by an earlier argument, and Assumptions E1 through E6 therefore hold if $\mathscr{S}$ is small enough and Equation 8.11 follows by Theorem 6.4 via Theorem 7.1*b*.

The treatment of the gain given by Equation 8.8*b* is virtually identical except for one small detail. The adaptive, "quick and dirty" gain of Equation 8.8*b* is not exactly of the form 7.48*c*. Whereas Equation 7.48*c* requires that the columns of $\mathbf{A}_k$ be of the form

$$\left[\sum_{j=1}^{\nu_{k+1}-1} \|\dot{\mathbf{F}}_j(\boldsymbol{\xi}_j)\|^2\right]^{-1} \dot{\mathbf{F}}_n(\boldsymbol{\xi}_n) \qquad (\nu_k \leq n < \nu_{k+1}),$$

where ξ_j takes values in $\mathscr{P}$ and depends on the observations through time j, the gain 8.8b is of the form

$$\left[\sum_{j=1}^{\nu_{k+1}-1} \|\dot{\mathbf{F}}_j(\mathbf{s}_k)\|^2\right]^{-1} \dot{\mathbf{F}}_n(\mathbf{s}_k) \qquad (\nu_k \leq n < \nu_{k+1}),$$

and $\mathbf{s}_k$ depends on the observations up through time $\nu_k - 1$. Nonetheless, the proof of Theorem 7.1 goes over word for word, and the same arguments used to establish the convergence of the recursion under the gain 8.8a can be applied verbatim to 8.8b.

The "linearized least-squares" gains of Equations 8.8c, d are treated similarly except that an additional assumption concerning the conditioning number of $\sum_{j=1}^{r} \dot{\mathbf{f}}_j(\boldsymbol{\theta}_0)\dot{\mathbf{f}}_j'(\boldsymbol{\theta}_0)$ is called for in order to meet Assumption F6 of Theorem 7.1.

In the very special case where $\mathbf{Z}_n = \mathbf{0}$ for every n and $r = p$, the regression problem reduces to that of finding the root of the equations

$$\begin{aligned} f_1(\boldsymbol{\theta}) &= Y_1 \\ f_2(\boldsymbol{\theta}) &= Y_2 \\ &\vdots \\ f_p(\boldsymbol{\theta}) &= Y_p. \end{aligned} \tag{8.15}$$

In the absence of noise, the vector "observations" are all the same:

$$\mathbf{Y}_k = \mathbf{Y} = \begin{bmatrix} Y_1 \\ \vdots \\ Y_p \end{bmatrix},$$

and so Equation 8.7 becomes

$$\mathbf{s}_{k+1} = [\mathbf{s}_k + \mathbf{A}_k(\mathbf{Y} - \mathbf{f}(\mathbf{s}_k))]_{\mathscr{P}}. \tag{8.16}$$

The preceding results show that

$$\lim_n \mathbf{s}_n = \boldsymbol{\theta},$$

where $\boldsymbol{\theta}$ is the root of Equation 8.15 in $\mathscr{P}$, provided that $\mathscr{P}$ is small enough. Actually, the rate of convergence can be speeded up considerably in the noiseless case by eliminating the damping factor $1/k$ from the gains of the type 8.8a, b, c, d. Convergence, then, follows from an easy extension of Theorem 6.1 to the case of batch processing.

8.2 Estimating the Initial State of a Linear System via Noisy Nonlinear Observations

Example 8.2. Suppose that a particle is moving back and forth along the x-axis, its position at time $t = k\tau$ being denoted by $x(k)$, and suppose that $x(k)$ satisfies the second-order difference equation

$$x(k + 1) + x(k - 1) = \sin k\alpha \qquad (k = 1, 2, \cdots),$$

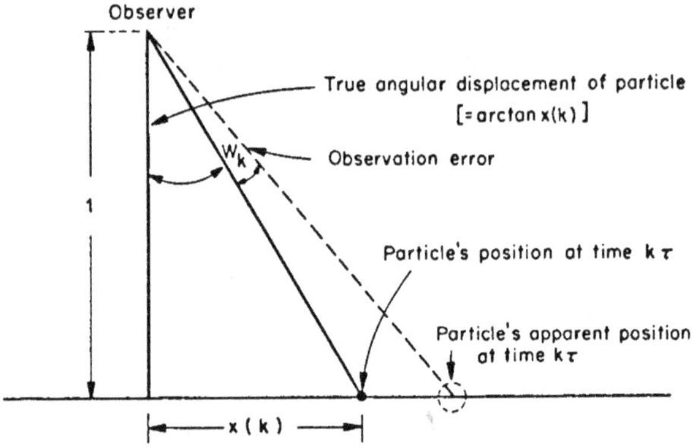

Figure 8.1 Observation geometry.

where α is known but the initial conditions

$$x(0) = \theta_1, \qquad x(1) = \theta_2$$

are not known. Suppose further that an observer located one unit away from the origin on a line passing through the origin and normal to the x-axis makes noisy observations on the angular displacement of the particle at the instants $k\tau$. (See Figure 8.1.)

Thus, the observations take the form

$$Y_k = \arctan x(k) + W_k \qquad (k = 1, 2, \cdots),$$

where it is assumed that the W_k are independent with zero-mean and common variances. We want to estimate

$$\boldsymbol{\theta} = \begin{bmatrix} \theta_1 \\ \theta_2 \end{bmatrix}$$

from the Y's. We proceed as follows:

The position of the particle can be written in closed form as

$$x(k; \boldsymbol{\theta}) = \varphi(k) + \mathbf{h}_k'\boldsymbol{\theta},$$

where

$$\varphi(k) = \sum_{n=1}^{k} \sin n\alpha \sin \frac{(k-n)\pi}{2},$$

and

$$\mathbf{h}_k = \begin{bmatrix} \cos k\pi/2 \\ \sin k\pi/2 \end{bmatrix}.$$

We assume that α is such that the system does not resonate and that $\boldsymbol{\theta}$ is known to lie within a sphere $\mathscr{S}$ of radius R, centered at the origin. In this case, there is a scalar C such that

$$\sup_{k, \boldsymbol{\theta} \in \mathscr{S}} |x(k; \boldsymbol{\theta})| \le C.$$

Our observations take the form

$$Y_k = F_k(\boldsymbol{\theta}) + W_k,$$

where

$$F_k(\boldsymbol{\xi}) = \arctan\,[\varphi(k) + \mathbf{h}_k'\boldsymbol{\xi}].$$

We estimate $\boldsymbol{\theta}$ by means of a scalar-observation recursion, truncated over $\mathscr{P}$:

$$\mathbf{s}_{n+1} = [\mathbf{s}_n + \mathbf{a}_n(Y_n - F_n(\mathbf{s}_n))]_{\mathscr{P}}.$$

The gains $\mathbf{a}_n$ can be chosen in a variety of ways. We will concentrate on "linearized least-squares" gains

$$\mathbf{a}_n = \left(\sum_{j=1}^{n} \dot{\mathbf{F}}_j\dot{\mathbf{F}}_j'\right)^{-1} \dot{\mathbf{F}}_n,$$

where the gradients $\dot{\mathbf{F}}_j$ can be evaluated either at some nominal value $\boldsymbol{\theta}_0$, (deterministic version) or at the then-most-recent estimate $\mathbf{s}_j$ (adaptive case).

In general,

$$\dot{\mathbf{F}}_n(\boldsymbol{\xi}) = [1 + (\varphi(n) + \mathbf{h}_n'\boldsymbol{\xi})^2]^{-1}\mathbf{h}_n;$$

therefore,

$$(1 + C^2)^{-1} \le \|\dot{\mathbf{F}}_n(\boldsymbol{\xi})\| \le 1,$$

and

$$\frac{\dot{\mathbf{F}}_n(\boldsymbol{\xi})}{\|\dot{\mathbf{F}}_n(\boldsymbol{\xi})\|} = \mathbf{h}_n = \begin{cases} (-1)^{n/2}\begin{pmatrix}1\\0\end{pmatrix} & \text{if } n \text{ is even,} \\[2ex] (-1)^{(n+1)/2}\begin{pmatrix}0\\1\end{pmatrix} & \text{if } n \text{ is odd.}\end{cases}$$

If we let

$$\alpha_n(\boldsymbol{\xi}) = [1 + (\varphi(n) + \mathbf{h}_n'\boldsymbol{\xi})^2]^{-1},$$

it is easy to see that the deterministic gains take the form

$$\begin{aligned}\mathbf{a}_{2n} &= \left(\sum_{j=1}^{2n} \alpha_j^2(\boldsymbol{\theta}_0)\mathbf{h}_j\mathbf{h}_j'\right)^{-1} \alpha_{2n}(\boldsymbol{\theta}_0)\mathbf{h}_{2n} \\ &= \left[\frac{\alpha_{2n}(\boldsymbol{\theta}_0)}{\sum_{j=1}^{n} \alpha_{2j}^2(\boldsymbol{\theta}_0)}\right]\mathbf{h}_{2n}, \\ \mathbf{a}_{2n+1} &= \left[\frac{\alpha_{2n+1}(\boldsymbol{\theta}_0)}{\sum_{j=1}^{n+1} \alpha_{2j-1}^2(\boldsymbol{\theta}_0)}\right]\mathbf{h}_{2n+1},\end{aligned}$$

while the adaptive gains take the form

$$\mathbf{a}_{2n} = \left[\frac{\alpha_{2n}(\mathbf{s}_{2n})}{\sum_{j=1}^{n} \alpha_{2j}^2(\mathbf{s}_{2j})}\right] \mathbf{h}_{2n},$$

$$\mathbf{a}_{2n+1} = \left[\frac{\alpha_{2n+1}(\mathbf{s}_{2n+1})}{\sum_{j=1}^{n+1} \alpha_{2j-1}^2(\mathbf{s}_{2j-1})}\right] \mathbf{h}_{2n+1}.$$

In either case, we have

$$\mathbf{a}_n/\|\mathbf{a}_n\| = \mathbf{h}_n,$$

and

$$0 < C_1/n \leq \|\mathbf{a}_n\| \leq C_2/n \qquad \text{uniformly in } \mathbf{a}_n\text{'s argument.}$$

It is now an easy matter to verify Assumptions E1 through E6′ of the conjectured theorem in Chapter 6. Assumptions E1, E2, and E6′ hold because

$$0 < C_3/n \leq \|\mathbf{a}_n\| \, \|\dot{\mathbf{F}}_n\| \leq C_2/n$$

uniformly in $\mathbf{a}_n$'s and $\dot{\mathbf{F}}_n$'s argument. Assumptions E3 and E4 hold with $\tau^2 = \frac{1}{2}$ and $\rho \leq C_2/C_3$ if we choose $\nu_k = 2k - 1$ $(k = 1, 2, \cdots)$. For then $p_k = 2$ for all k and

$$\frac{1}{p_k} \lambda_{\min}\left(\sum_{j=2k-1}^{2k} \frac{\dot{\mathbf{F}}_j \dot{\mathbf{F}}_j'}{\|\dot{\mathbf{F}}_j\|^2}\right) = \tfrac{1}{2}\lambda_{\min}(I) = \tfrac{1}{2},$$

while

$$\max_{j \in J_k} \|\mathbf{a}_j\| \, \|\dot{\mathbf{F}}_j\| \leq C_2/2k - 1, \qquad \text{and} \qquad \min_{j \in J_k} \|\mathbf{a}_j\| \, \|\dot{\mathbf{F}}_j\| \geq C_3/2k.$$

Finally, Assumption E5 holds, since

$$1 = \frac{\mathbf{a}_n' \dot{\mathbf{F}}_n}{\|\mathbf{a}_n\| \, \|\dot{\mathbf{F}}_n\|} > \sqrt{\frac{1 - \tau^2}{1 - \tau^2 + \tau^2/\rho^2}}.$$

The same results can be obtained if a batch-processing recursion (with linearized least-squares gains) truncated over $\mathscr{P}$ is used. Theorem 6.4 can then be applied.

8.3 Estimating Input Amplitude Through an Unknown Saturating Amplifier

Example 8.3. Amplifiers are, ideally, memoryless linear devices. *Real* amplifiers only approximate this performance. They are practically memoryless but linear only over a certain range of inputs. Typically, as

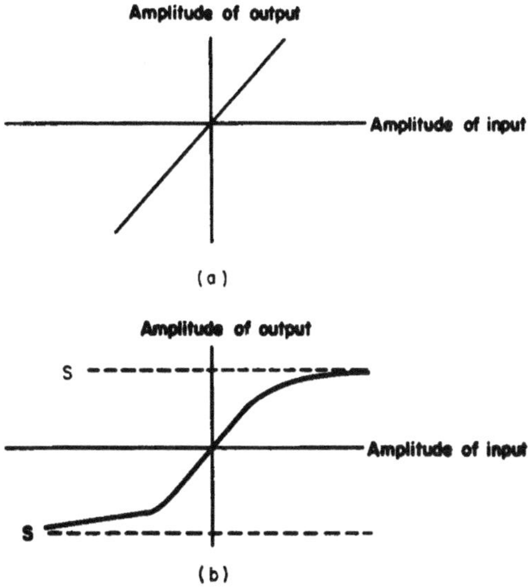

Figure 8.2 *(a)* **Performance of an ideal amplifier.** *(b)* **Performance of a real amplifier.**

the input amplitude increases, the amplifier saturates. (See Figures 8.2*a* and 8.2*b*.) A model that is frequently used to describe the input–output relationship of a saturating amplifier states that

$$Y_{\text{out}}(t) = \frac{2S}{\pi} \arctan\left(\frac{\pi A}{2S} Y_{\text{in}}(t)\right).$$

Such an amplifier has the property that

$$Y_{\text{out}}(t) \approx A Y_{\text{in}}(t) \qquad \text{if} \quad A Y_{\text{in}}(t) \ll S,$$

and

$$Y_{\text{out}}(t) \approx S \qquad \text{if} \quad A Y_{\text{in}}(t) \gg S.$$

Here A is called the amplification factor, and S is called the saturation level of the amplifier.

Suppose that a sinusoid $B \sin(2\pi f t + \Psi)$ of known frequency and phase but unknown amplitude is passed through an amplifier whose saturation level S is unknown, and suppose that the output is observed in the presence of wide-band noise. In other words, we observe

$$Y(t) = \frac{2S}{\pi} \arctan\left[\left(\frac{\pi A B}{2S}\right) \sin(2\pi f t + \Psi)\right] + Z(t).$$

On the basis of these observations, it is possible to estimate S and AB. (If A is known, we can deduce B. Otherwise we can only estimate their product.)

For notational convenience, we will sample $Y(t)$ at times

$$t_k = \begin{cases} \left(\dfrac{k\pi}{2} - \Psi\right)\Big/2\pi f & \text{if } k \text{ is odd,} \\ \left(\dfrac{2k+1}{4}\pi - \Psi\right)\Big/2\pi f & \text{if } k \text{ is even.} \end{cases}$$

If we set

$$g_k = \sin(2\pi f t_k + \Psi) = \begin{cases} (-1)^{(k-1)/2}, & k \text{ odd,} \\ (-1)^{k/2}/\sqrt{2}, & k \text{ even,} \end{cases}$$

$$Z_k = Z(t_k), \qquad Y_k = Y(t_k), \qquad \theta_1 = \frac{2S}{\pi}, \qquad \theta_2 = \frac{\pi AB}{2S},$$

and

$$F_k(\boldsymbol{\theta}) = \theta_1 \arctan \theta_2 g_k, \tag{8.17}$$

then we can write

$$Y_k = F_k(\boldsymbol{\theta}) + Z_k \qquad (k = 1, 2, \cdots).$$

We assume that

$$0 < \alpha_1 \le \theta_1 \le \beta_1 < \infty, \qquad \text{and} \qquad 0 < \alpha_2 \le \theta_2 \le \beta_2 < \infty,$$

and estimate $\boldsymbol{\theta}$ via the batch-processing recursion

$$\mathbf{s}_{n+1} = [\mathbf{s}_n + \mathbf{A}_n(\mathbf{Y}_n - \mathbf{f}_n(\mathbf{s}_n))]_{\mathscr{P}}, \qquad \mathbf{s}_1 = \boldsymbol{\theta}_0 \in \mathscr{P},$$

where $\mathscr{P}$ is the rectangular parallelepiped, $[\alpha_1, \beta_1] \otimes [\alpha_2, \beta_2]$,

$$\mathbf{Y}_n = \begin{bmatrix} Y_{2n-1} \\ Y_{2n} \end{bmatrix}, \qquad \mathbf{f}_n(\cdot) = \begin{bmatrix} F_{2n-1}(\cdot) \\ F_{2n}(\cdot) \end{bmatrix},$$

and

$$\mathbf{A}_n = \left(\sum_{j=1}^{2n} \|\dot{\mathbf{F}}_j(\boldsymbol{\theta}_0)\|^2\right)^{-1} (\dot{\mathbf{F}}_{2n-1}(\boldsymbol{\theta}_0), \dot{\mathbf{F}}_{2n}(\boldsymbol{\theta}_0)), \tag{8.18a}$$

or

$$\mathbf{A}_n = \left(\sum_{j=1}^{2n} \dot{\mathbf{F}}_j(\boldsymbol{\theta}_0)\dot{\mathbf{F}}_j'(\boldsymbol{\theta}_0)\right)^{-1} (\dot{\mathbf{F}}_{2n-1}(\boldsymbol{\theta}_0), \dot{\mathbf{F}}_{2n}(\boldsymbol{\theta}_0)). \tag{8.18b}$$

The conditions of Theorem 7.1 are dealt with as follows: The gradient of the regression function is

$$\dot{\mathbf{F}}_k(\boldsymbol{\theta}) = \begin{bmatrix} \arctan \theta_2 g_k \\ \dfrac{\theta_1}{1 + (\theta_2 g_k)^2} \end{bmatrix}, \tag{8.19}$$

and the matrix of F_k's mixed partials, the first column evaluated at

$\mathbf{x} = \begin{pmatrix} x_1 \\ x_2 \end{pmatrix}$, the second at $\mathbf{z} = \begin{pmatrix} z_1 \\ z_2 \end{pmatrix}$, is given by

$$\mathbf{G}_k = \begin{bmatrix} 0 & [1 + (z_2 g_k)^2]^{-1} \\ [1 + (x_2 g_k)^2]^{-1} & -2z_1 z_2 g_k [1 + (z_2 g_k)^2]^{-2} \end{bmatrix}.$$

Given the existing assumptions, the norm of $\dot{\mathbf{F}}_k(\cdot)$ is uniformly (in k and $\boldsymbol{\theta}$) bounded above and away from zero, and the columns of $\mathbf{G}_k$ are uniformly bounded. This establishes Assumptions F1 through F4. To establish Assumption F5, we choose

$$\nu_k = 2k - 1 \qquad (k = 1, 2, \cdots).$$

The norms $\|\dot{\mathbf{F}}_k(\mathbf{x})\|$ are uniformly bounded (in $\mathbf{x}$ and k); therefore, it suffices to show that for some $\mathbf{x} \in \mathscr{P}$ and some $\delta > 0$,

$$\lambda_{\min} (\dot{\mathbf{F}}_{2k-1}(\mathbf{x})\dot{\mathbf{F}}'_{2k-1}(\mathbf{x}) + \dot{\mathbf{F}}_{2k}(\mathbf{x})\dot{\mathbf{F}}'_{2k}(\mathbf{x})) \geq \delta \tag{8.20}$$

for all k. Equation 8.20 will follow if we show that $\dot{\mathbf{F}}_{2k}$ and $\dot{\mathbf{F}}_{2k-1}$ are linearly independent for every k. Since $\dot{\mathbf{F}}_{2k-1} = \dot{\mathbf{F}}_{2k+3}$ and $\dot{\mathbf{F}}_{2k} = \dot{\mathbf{F}}_{2k+4}$, it suffices to do so for $k = 1$. Let us assume the contrary:

$$\dot{\mathbf{F}}_1(\mathbf{x}) = a\dot{\mathbf{F}}_2(\mathbf{x}) \qquad \text{for some } \mathbf{x} \in \mathscr{P} \qquad \text{and} \quad \text{some nonzero } a.$$

Then, since $g_1 = 1$ and $g_2 = -1\sqrt{2}$, we have, by Equation 8.19,

$$\arctan(x_2) = a \arctan(-x_2/\sqrt{2}),$$

and

$$(1 + x_2^2)^{-1} = a(1 + x_2^2/2)^{-1}.$$

Multiplying the first relation by the second, we obtain

$$\frac{\arctan(x_2)}{1 + (x_2)^2} = \frac{\arctan(-x_2/\sqrt{2})}{1 + (x_2/\sqrt{2})^2}. \tag{8.21}$$

By assumption, we see that $x_2 > 0$; therefore, the left-hand side is positive, while the right-hand side is negative. This establishes a contradiction. Therefore, $\dot{\mathbf{F}}_{2k}$ and $\dot{\mathbf{F}}_{2k-1}$ are indeed linearly independent for every k, and Assumption F5 holds. If $(\beta_i - \alpha_i)$ $(i = 1, 2)$ are small, $\mathbf{s}_n$ converges to $\boldsymbol{\theta}$ in the mean square if the gain of Equation 8.18*a* is used. If Equation 8.18*b* is used, the additional restriction on the conditioning number of the matrices

$$\dot{\mathbf{F}}_{2k-1}(\mathbf{x})\dot{\mathbf{F}}'_{2k-1}(\mathbf{x}) + \dot{\mathbf{F}}_{2k}(\mathbf{x})\dot{\mathbf{F}}'_{2k}(x) \qquad (k = 1, 2)$$

must be satisfied.

In this treatment of the example, we chose batches of two observations each. If we choose batches of size four, each "observation" is of the form

$$\mathbf{Y}_k^* = \begin{bmatrix} Y_{4k-3} \\ Y_{4k-2} \\ Y_{4k-1} \\ Y_{4k} \end{bmatrix} = \mathbf{f}(\boldsymbol{\theta}) + \mathbf{Z}_k^* \qquad (k = 1, 2, \cdots)$$

where now

$$\mathbf{f}(\boldsymbol{\theta}) = \begin{bmatrix} F_{4k-3}(\boldsymbol{\theta}) \\ F_{4k-2}(\boldsymbol{\theta}) \\ F_{4k-1}(\boldsymbol{\theta}) \\ F_{4k}(\boldsymbol{\theta}) \end{bmatrix}$$

does not depend upon k, since $F_k(\cdot) = F_{k+4}(\cdot)$ $(k = 1, 2, \cdots)$ and $\mathbf{Z}_k^*$ is defined in the obvious way. The recursion becomes

$$\mathbf{s}_{n+1}^* = [\mathbf{s}_n^* + \mathbf{A}_n^*(\mathbf{Y}_n^* - \mathbf{f}(\mathbf{s}_n^*))]_{\mathscr{P}},$$

where

$$\mathbf{A}_n^* = \frac{1}{n}\left(\sum_{k=1}^{4} \|\dot{\mathbf{F}}_k(\boldsymbol{\theta}_0)\|^2\right)^{-1} (\dot{\mathbf{F}}_1(\boldsymbol{\theta}_0), \dot{\mathbf{F}}_2(\boldsymbol{\theta}_0), \dot{\mathbf{F}}_3(\boldsymbol{\theta}_0), \dot{\mathbf{F}}_4(\boldsymbol{\theta}_0)),$$

or

$$\mathbf{A}_n^* = \frac{1}{n}\left(\sum_{k=1}^{4} \dot{\mathbf{F}}_k(\boldsymbol{\theta}_0)\dot{\mathbf{F}}_k'(\boldsymbol{\theta}_0)\right)^{-1} (\dot{\mathbf{F}}_1(\boldsymbol{\theta}_0), \dot{\mathbf{F}}_2(\boldsymbol{\theta}_0), \dot{\mathbf{F}}_3(\boldsymbol{\theta}_0), \dot{\mathbf{F}}_4(\boldsymbol{\theta}_0)).$$

Assumptions F1 through F4 of Theorem 7.1 follow from previous arguments, while Assumption F5 follows from the fact that

$$\lambda_{\min}\left(\sum_{j=1}^{4} \dot{\mathbf{F}}_j(\boldsymbol{\theta}_0)\dot{\mathbf{F}}_j'(\boldsymbol{\theta}_0)\right) \geq \lambda_{\min}\left(\sum_{j=1}^{2} \dot{\mathbf{F}}_j(\boldsymbol{\theta}_0)\dot{\mathbf{F}}_j'(\boldsymbol{\theta}_0)\right) \geq \delta > 0.$$

As usual, a restriction on the conditioning number of $\sum_{j=1}^{4} \dot{F}_j(\theta_0)\dot{F}_j'(\theta_0)$ must be met if the second ("linearized least-squares") type is to be used.

8.4 Estimating the Parameters of a Time-Invariant Linear System

Example 8.4. Here we consider recursive estimation of the parameters defining a stable time-invariant linear system when it is driven by appropriate inputs. The output is observed in the presence of additive, but not necessarily white, noise. We treat both continuous and discrete time systems. When the continuous output is sampled at regular intervals, as is usually done in practice, the two estimation procedures are very much the same. As we will see, however, there is a single difference which is important from the computational point of view.

Our results on the asymptotic behavior of the estimates are quite complete (strong and mean-square convergence plus asymptotic normality). Although only an indication of proof is given for some results, each can be established rigorously under the stated conditions.

Consider an output $x(t)$ that satisfies a stable pth-order linear differential equation

$$\sum_{j=0}^{p} \theta_j \frac{d^j x(t)}{dt^j} = g(t) \qquad (-\infty < t < \infty), \tag{8.22C}$$

or a stable pth-order linear difference equation, which we write as

$$\sum_{j=0}^{p} \theta_j x(t-j) = g(t) \qquad (t = \cdots, -1, 0, +1, \cdots). \tag{8.22D}$$

In either case, if

$$g(t) = \cos \omega t, \tag{8.23}$$

the steady-state output takes the form

$$x(t) = A \cos \omega t + B \sin \omega t, \tag{8.24}$$

where A and B depend nonlinearly on the θ's.

To exhibit this dependence in the continuous case, we can compute the even- and odd-ordered time derivatives of 8.24 and substitute into 8.22C. Letting $[x]$ denote the integral part of x, we find that

$$\sum_{j=0}^{p} \theta_j \frac{d^j x(t)}{dt^j} = (A \cos \omega t + B \sin \omega t) \sum_{j=0}^{[p/2]} \theta_{2j}(-1)^j \omega^{2j}$$
$$+ (A \sin \omega t - B \cos \omega t) \sum_{j=0}^{[(p-1)/2]} \theta_{2j+1}(-1)^{j+1} \omega^{2j+1}.$$

If this is to equal 8.23 for all $t > 0$, the coefficient of $\cos \omega t$ **must** be unity and that of $\sin \omega t$ must be zero. As a result, we have

$$\frac{A}{A^2 + B^2} = \alpha, \qquad \frac{B}{A^2 + B^2} = \beta, \tag{8.25}$$

where α and β are linearly related to the unknown parameters via

$$\alpha = \sum_{j=0}^{[p/2]} \theta_{2j}(-1)^j \omega^{2j}, \qquad \beta = \sum_{j=0}^{[(p-1)/2]} \theta_{2j+1}(-1)^j \omega^{2j+1}. \tag{8.26C}$$

In the discrete case, Equation 8.25 is again easily shown to be a valid relation, after we redefine α and β by

$$\alpha = \sum_{j=0}^{p} \theta_j \cos j\omega, \qquad \beta = -\sum_{j=0}^{p} \theta_j \sin j\omega. \tag{8.26D}$$

We note that Equation 8.25 holds reciprocally, that is, with A interchanged with α and B with β, thereby making explicit the nonlinear dependence of A and B on the θ's.

For the sake of convenience we are going to restrict attention to the case where the number of unknown parameters is even, that is, where

$$p = 2q + 1 \tag{8.27}$$

for some integer $q \geq 0$. The modifications required when $p + 1$ is odd will be clear.

To estimate the parameters in the continuous-time case, we will take as our input

$$g(t) = \sum_{k=0}^{q} \cos \lambda_k t \qquad (-\infty < t < \infty), \tag{8.28}$$

where the λ's are distinct positive angular frequencies to be chosen so that

$$\lambda_k \pm \lambda_j \neq \text{a multiple of } \pi \qquad \text{if} \quad k \neq j.$$

The superposition principle immediately allows us to write the steady-state output as

$$F_c(t; \boldsymbol{\theta}) = \sum_{k=0}^{q} (A_k \cos \lambda_k t + B_k \sin \lambda_k t) \qquad (t > 0), \tag{8.29}$$

where the $2(q + 1)$-coefficients, A_k and B_k, are related to the θ's via Equations 8.25 and 8.26C after setting $\omega = \lambda$ and affixing the subscript $k = 0, 1, \cdots, q$ to each of A, B, α, β, and λ. In view of Equation 8.27,

$[p/2]$ and $[(p-1)/2]$ are both equal to q; therefore (with "e" for even and "o" for odd), we have

$$\boldsymbol{\alpha} = \begin{bmatrix} \alpha_0 \\ \alpha_1 \\ \vdots \\ \alpha_q \end{bmatrix} = \boldsymbol{\Lambda}_e \boldsymbol{\theta}_e, \qquad \boldsymbol{\beta} = \begin{bmatrix} \beta_0 \\ \beta_1 \\ \vdots \\ \beta_q \end{bmatrix} = \boldsymbol{\Lambda}_o \boldsymbol{\theta}_o, \tag{8.30}$$

where

$$\boldsymbol{\theta}_e = \begin{bmatrix} \theta_0 \\ \theta_2 \\ \vdots \\ \theta_{2q} \end{bmatrix} \qquad \boldsymbol{\theta}_o = \begin{bmatrix} \theta_1 \\ \theta_3 \\ \vdots \\ \theta_{2q+1} \end{bmatrix} \tag{8.31a}$$

$$\boldsymbol{\Lambda}_e = \begin{bmatrix} 1 & -\lambda_0^2 & \lambda_0^4 & & (-1)^q \lambda_0^{2q} \\ 1 & -\lambda_1^2 & \lambda_1^4 & & (-1)^q \lambda_1^{2q} \\ \vdots & & & & \\ 1 & -\lambda_q^2 & \lambda_q^4 & & (-1)^q \lambda_q^{2q} \end{bmatrix},$$

$$\boldsymbol{\Lambda}_o = \begin{bmatrix} \lambda_0 & -\lambda_0^3 & \lambda_0^5 & & (-1)^q \lambda_0^{2q+1} \\ \lambda_1 & -\lambda_1^3 & \lambda_1^5 & & (-1)^q \lambda_1^{2q+1} \\ \vdots & & & & \vdots \\ \lambda_q & -\lambda_q^3 & \lambda_q^5 & & (-1)^q \lambda_q^{2q+1} \end{bmatrix} \tag{8.31b}$$

In the discrete-time case, we take as our input

$$g(t) = \frac{1}{\sqrt{2}} + \sum_{k=1}^{q} \cos \omega_k t + \frac{(-1)^t}{\sqrt{2}} \qquad (t = \cdots, -1, 0, +1, \cdots), \tag{8.32}$$

where

$$0 < \omega_1 < \omega_2 < \cdots < \omega_q < \pi.$$

The corresponding output is

$$F_d(t; \boldsymbol{\theta}) = \frac{A_0}{\sqrt{2}} + \sum_{k=1}^{q} (A_k \cos \omega_k t + B_k \sin \omega_k t) + \frac{A_{q+1}}{\sqrt{2}} (-1)^t \qquad (t = 1, 2, \cdots), \tag{8.33}$$

where $A_0, A_1, B_1, \cdots, A_q, B_q, A_{q+1}$ are related to $\theta_0, \theta_1, \cdots, \theta_{2q+1}$ via Equations 8.25 and 8.26D, after we affix $k = 0, 1, \cdots, q + 1$ to each of A, B, α, β, and ω and set

$$\omega_0 = 0, \qquad \omega_{q+1} = \pi$$

(so that $B_k = \beta_k = 0$ for $k = 0$ and $q + 1$). Thus, we have

$$\boldsymbol{\Upsilon} = \begin{bmatrix} \alpha_0 \\ \alpha_1 \\ \beta_1 \\ \vdots \\ \alpha_q \\ \beta_q \\ \alpha_{q+1} \end{bmatrix} = \boldsymbol{\Omega} \begin{bmatrix} \theta_0 \\ \theta_1 \\ \vdots \\ \theta_{2q+1} \end{bmatrix} = \boldsymbol{\Omega\theta}, \tag{8.34}$$

where

$$\boldsymbol{\Omega} = \begin{bmatrix} 1 & 1 & 1 & & 1 \\ 1 & \cos \omega_1 & \cos 2\omega_1 & & \cos (2q+1)\omega_1 \\ 0 & -\sin \omega_1 & -\sin 2\omega_1 & & -\sin (2q+1)\omega_1 \\ \vdots & & & & \\ 1 & \cos \omega_q & \cos 2\omega_q & & \cos (2q+1)\omega_q \\ 0 & -\sin \omega_q & -\sin 2\omega_q & & -\sin (2q+1)\omega_q \\ 1 & -1 & 1 & & -1 \end{bmatrix} \tag{8.35}$$

is $2(q+1)$ by $2(q+1)$, while each of $\boldsymbol{\Lambda}_e$ and $\boldsymbol{\Lambda}_o$ is $q+1$ by $q+1$.

Without loss of generality, we take the positive integers as the sampling instants for a continuous output. (For a constant increment $\tau \neq 1$, the restrictions are placed on $\lambda_k \tau$ rather than on λ_k.) Our observations, in both the sampled continuous and discrete cases, are then of the form

$$Y_t = \mathbf{h}_t'\boldsymbol{\xi} + Z_t \qquad (t = 1, 2, \cdots), \tag{8.36}$$

where, according to Equations 8.29 and 8.33,

$$\boldsymbol{\xi} = \begin{bmatrix} A_0 \\ B_0 \\ A_1 \\ B_1 \\ \vdots \\ A_q \\ B_q \end{bmatrix} \qquad \left.\begin{aligned} \frac{A_k}{A_k^2 + B_k^2} &= \alpha_k \\ \frac{B_k}{A_k^2 + B_k^2} &= \beta_k \end{aligned}\right\} \text{ which are related to } \boldsymbol{\theta} \text{ via the linear equations 8.30 with } k = 0, 1, \cdots, q, \tag{8.37C}$$

or

$$\xi = \begin{bmatrix} A_0 \\ A_1 \\ B_1 \\ \vdots \\ A_q \\ B_q \\ A_{q+1} \end{bmatrix} \quad \left.\begin{aligned} \frac{A_k}{A_k^2 + B_k^2} &= \alpha_k \\ \frac{B_k}{A_k^2 + B_k^2} &= \beta_k \end{aligned}\right\} \text{ which are related to } \boldsymbol{\theta} \text{ via the linear equations 8.34 with } k = 0, 1, \cdots, q+1. \tag{8.37D}$$

and

$$\mathbf{h}_t' = [\cos \lambda_0 t, \sin \lambda_0 t, \cos \lambda_1 t, \sin \lambda_1 t, \cdots, \cos \lambda_q t, \sin \lambda_q t], \tag{8.38C}$$

or

$$\mathbf{h}_t' = \left[\frac{1}{\sqrt{2}}, \cos \omega_1 t, \sin \omega_1 t, \cdots, \cos \omega_q t, \sin \omega_q t, \frac{(-1)^t}{\sqrt{2}}\right]. \tag{8.38D}$$

We will assume that $\{Z_t : t = 1, 2, \cdots\}$ is some zero-mean, second order stationary process. Initially, our only restriction on the unknown covariance sequence is that

$$\sigma_h = \mathscr{E} Z_t Z_{t+|h|} = O\left(\frac{1}{h^\varepsilon}\right), \tag{8.39}$$

as $h \to \infty$ for some ε, $0 < \varepsilon < 1$ (and thus the noise process need not possess a spectral density).

We are now in a position to write down the procedure for estimating $\boldsymbol{\theta}$. We affix an n to a parameter symbol to denote an estimate of the parameter based on the first n observations. Thus, ξ_n is a vector-valued function of $Y_1, Y_2, \cdots, Y_n$ which estimates ξ, $A_{k,n}$ estimates the component A_k, and so on.

Step 1. Recursively estimate the $(p + 1)$-dimensional parameter ξ by naïve least squares:

$$\xi_n = \xi_{n-1} + \mathbf{B}_n \mathbf{h}_n (Y_n - \mathbf{h}_n' \xi_{n-1}),$$

where

$$\mathbf{B}_n = \mathbf{B}_{n-1} - \frac{(\mathbf{B}_{n-1}\mathbf{h}_n)(\mathbf{B}_{n-1}\mathbf{h}_n)'}{1 + \mathbf{h}_n' \mathbf{B}_{n-1} \mathbf{h}_n} \qquad (n \geq 2q + 3),$$

initialized by

$$\mathbf{B}_{2q+2} = \left(\sum_{t=1}^{2q+2} \mathbf{h}_t \mathbf{h}_t'\right)^{-1} \qquad \xi_{2q+2} = \mathbf{B}_{2q+2} \sum_{t=1}^{2q+2} \mathbf{h}_t Y_t.$$

(See the discussion leading to Equation 7.15; here we have decreased the iteration index n by one.)

Step 2. Assume that a finite number Δ is known for which

$$\max_{j=0,1,\cdots,2q+1} |\theta_j| < \Delta.$$

Compute

$$\alpha_{k,n} = \left[\frac{A_{k,n}}{A_{k,n}^2 + B_{k,n}^2}\right]_{I_k},$$

$$\beta_{k,n} = \left[\frac{B_{k,n}}{A_{k,n}^2 + B_{k,n}^2}\right]_{I_k}$$

where in the sampled continuous case

$$I_k = \{x : |x| \leq (q+1)\Delta \max_{j=0,1,\cdots,q} |\lambda_k|^{2j+1}\}$$

and in the discrete case

$$I_k = \{x : |x| \leq 2(q+1)\Delta\} \qquad \text{for all } k = 0, 1, \cdots, q+1.$$

Step 3. In the sampled continuous case, form the vectors $\boldsymbol{\alpha}_n$ and $\boldsymbol{\beta}_n$ whose components are, respectively, $\alpha_{k,n}$ and $\beta_{k,n}$ ($k = 1, \cdots, q$). Then estimate the even-numbered components of $\boldsymbol{\theta}$ by $\boldsymbol{\theta}_{e,n}$ and the odd-numbered components by $\boldsymbol{\theta}_{o,n}$, where

$$\boldsymbol{\theta}_{e,n} = \boldsymbol{\Lambda}_e^{-1}\boldsymbol{\alpha}_n, \qquad \text{and} \qquad \boldsymbol{\theta}_{o,n} = \boldsymbol{\Lambda}_o^{-1}\boldsymbol{\beta}_n.$$

In the discrete case, form the vector $\boldsymbol{\gamma}_n$ whose components are given by Equation 8.35 with the α_k's and β_k's replaced throughout by their respective estimates $\alpha_{k,n}$ and $\beta_{k,n}$. Then estimate $\boldsymbol{\theta}$ by

$$\boldsymbol{\theta}_n = \boldsymbol{\Omega}^{-1}\boldsymbol{\gamma}_n.$$

We first show that Equation 8.39 is sufficient for *probability-one and mean-square convergence* of $\boldsymbol{\theta}_n$ to $\boldsymbol{\theta}$ as $n \to \infty$: Let $\mathbf{y}_n$ and $\mathbf{z}_n$, respectively, be the vector of the first n observations and first n noise realizations, and let

$$\mathbf{H}_n = [\mathbf{h}_1, \mathbf{h}_2, \cdots, \mathbf{h}_n]$$

be the $2(q+1)$ by n matrix whose columns are given either by Equation 8.38C or 8.38D. Equations 8.36 and the closed-form expression for $\boldsymbol{\xi}_n$ combine to give

$$\mathbf{B}_n^{-1}(\boldsymbol{\xi}_n - \boldsymbol{\xi}) = \mathbf{H}_n\mathbf{z}_n, \tag{8.40}$$

where

$$\mathbf{B}_n^{-1} = \sum_{t=1}^{n} \mathbf{h}_t\mathbf{h}_t', \qquad n > 2(q+1).$$

From the identities (Knopp, 1947, p. 480)

$$\left.\begin{aligned}\sum_{t=1}^{n} \cos 2\lambda t &= \frac{\sin n\lambda}{\sin \lambda} \cos (n+1)\lambda \\ \sum_{t=1}^{n} \sin 2\lambda t &= \frac{\sin n\lambda}{\sin \lambda} \sin (n+1)\lambda\end{aligned}\right\} \quad (\lambda \neq \text{a multiple of } \pi), \tag{8.41}$$

it follows for n tending to infinity that

$$\left.\begin{aligned}\frac{1}{n}\sum_{t=1}^{n} \cos^2 \lambda t \\ \frac{1}{n}\sum_{t=1}^{n} \sin^2 \lambda t\end{aligned}\right\} = \tfrac{1}{2} + O\left(\frac{1}{n}\right) \quad (\lambda \neq \text{a multiple of } \pi),$$

$$\left.\begin{aligned}\frac{1}{n}\sum_{t=1}^{n} \cos \lambda_1 t \cos \lambda_2 t \\ \frac{1}{n}\sum_{t=1}^{n} \sin \lambda_1 t \sin \lambda_2 t\end{aligned}\right\} = O\left(\frac{1}{n}\right) \quad (\lambda_1 \neq \lambda_2) \tag{8.42}$$

$$\frac{1}{n}\sum_{t=1}^{n} \cos \lambda_1 t \sin \lambda_2 t = O\left(\frac{1}{n}\right) \quad (\text{all } \lambda_1, \lambda_2).$$

Consequently, for the **h**-vectors defined either by Equation 8.38C or 8.38D, we have $(2/n)\mathbf{B}_n^{-1} = \mathbf{I} + (1/n)\mathbf{E}_n$ for some matrix $\mathbf{E}_n$, whose elements remain uniformly bounded as $n \to \infty$. From Equation 8.40, therefore,

$$\left(\mathbf{I} + \frac{1}{n}\mathbf{E}_n\right)(\boldsymbol{\xi}_n - \boldsymbol{\xi}) = \frac{2}{n}\mathbf{H}_n\mathbf{z}_n, \tag{8.43}$$

and $\boldsymbol{\xi}_n$ will converge to $\boldsymbol{\xi}$ in the same probabilistic sense that the $2(q+1)$-vector on the right-hand side converges to the zero vector. Ignoring the multiplication factor, each entry is of the form

$$\frac{1}{n}\sum_{t=1}^{n} c_t Z_t = \bar{Z}_n,$$

where c_t is either a cosine or sine. Using our assumption, 8.39, we find that

$$\mathscr{E}|\bar{Z}_n c_n Z_n| \leq \frac{1}{n}\sum_{t=1}^{n-1} |\sigma_t| = O\left(\frac{1}{n^{\varepsilon}}\right)$$

as $n \to \infty$. It follows from a Law of Large Numbers for centered dependent random variables (Parzen, 1960, p. 419) that $\bar{Z}_n$ tends to zero with probability one as $n \to \infty$.

In Step 2, therefore, each $A_{k,n}$ and $B_{k,n}$ is a strongly consistent estimate of the corresponding values of A_k and B_k. Since strong convergence is retained through continuous transformations, we have

$$\alpha_{k,n} \xrightarrow{\text{a.s.}} [\alpha_k]_{I_k} = \alpha_k$$

as $n \to \infty$. But $\alpha_{k,n}$ is uniformly bounded in n, so it follows that $\alpha_{k,n}$ converges to α_k in every mean (see the introductory material of Chapter 4). The same clearly applies to the convergence of $\beta_{k,n}$ to β_k. Consequently, the components of the vectors $\boldsymbol{\alpha}_n$ and $\boldsymbol{\beta}_n$ or $\boldsymbol{\gamma}_n$ in Step 3 converge with probability one and in mean square to the corresponding components of $\boldsymbol{\alpha}$ and $\boldsymbol{\beta}$ or $\boldsymbol{\gamma}$. The $\boldsymbol{\theta}$ estimates are obtained by applying linear transformations to these vector estimates. Consequently, both types of convergence are maintained.

It is an easy matter to establish *asymptotic normality* of our estimates when we restrict attention to independent noise processes with

$$\mathscr{E}Z_t^2 = \sigma^2 \qquad \text{and} \qquad \sup_{t=1,2,\cdots} \mathscr{E}|Z_t|^{2+\varepsilon} < \infty. \tag{8.44}$$

From Equation 8.43, we find that

$$\sqrt{n}(\boldsymbol{\xi}_n - \boldsymbol{\xi}) \sim \frac{2}{\sqrt{n}} \mathbf{H}_n \mathbf{z}_n = \frac{2}{\sqrt{n}} \sum_{t=1}^{n} \mathbf{h}_t Z_t$$

as $n \to \infty$. From Equation 8.44 and Liapounov's Central Limit Theorem (Loève, 1960, p. 275), it can be shown that the vector on the right-hand side tends to normality with zero-mean vector and covariance matrix equal to

$$\begin{aligned}\lim_n \frac{4}{n} \mathscr{E}\left(\sum_{t=1}^{n} \mathbf{h}_t Z_t\right)\left(\sum_{s=1}^{n} \mathbf{h}_s Z_s\right)' &= \lim_n \frac{4}{n} \sum_{t=1}^{n} \sum_{s=1}^{n} \mathbf{h}_t \mathbf{h}_s' \mathscr{E} Z_t Z_s \\ &= \lim_n \frac{4\sigma^2}{n} \mathbf{B}_n^{-1} = 2\sigma^2 \mathbf{I},\end{aligned} \tag{8.45}$$

where $\mathbf{I}$ is the $(p + 1)$-by-$(p + 1)$ identity. Equation 8.45 is true in both the sampled continuous and discrete cases. It is now easy to establish the asymptotic normality of the $\boldsymbol{\theta}$ estimates.

Turning first to the relations in Equation 8.37C, we consider the matrix of partial derivatives

$$\frac{\partial(\alpha_0, \alpha_1, \cdots, \alpha_q, \beta_0, \beta_1, \cdots, \beta_q)}{\partial(A_0, B_0, A_1, B_1, \cdots, A_q, B_q)} = \mathbf{J}$$

evaluated at the true parameter values. Since α_k and β_k are interior points of I_k, the vector

$$\begin{bmatrix} \boldsymbol{\alpha}_n \\ \boldsymbol{\beta}_n \end{bmatrix}$$

resulting from Step 2 tends, after appropriate standardization, to zero-mean normality with covariance matrix $2\sigma^2\mathbf{J}'\mathbf{J}$ (by the vector version of the "delta method" used in the proof of Theorem 5.1). After computing the derivatives, we find that $\mathbf{J}'\mathbf{J}$ is a diagonal matrix, and that the second $q + 1$ diagonal entries are identical to the first $q + 1$, namely, $(\alpha_0^2 + \beta_0^2)^2, (\alpha_1^2 + \beta_1^2)^2, \cdots, (\alpha_q^2 + \beta_q^2)^2$.

If we set

$$\rho_k = \alpha_k^2 + \beta_k^2, \tag{8.46}$$

it follows that $\sqrt{n}(\boldsymbol{\alpha}_n - \boldsymbol{\alpha})$ and $\sqrt{n}(\boldsymbol{\beta}_n - \boldsymbol{\beta})$ are asymptotically independent and identically distributed as a $(q + 1)$-dimensional normal random variable with $\mathbf{0}$ mean and covariance matrix $2\sigma^2\mathbf{P}^2$, where

$$\mathbf{P} = \text{diag}\,[\rho_0, \rho_1, \cdots, \rho_q].$$

Consequently, for the even and odd components of the estimate of $\boldsymbol{\theta}$ in Step 3, there results

$$\begin{aligned} \sqrt{n}(\boldsymbol{\theta}_{e,n} - \boldsymbol{\theta}_e) &\sim N(0, 2\sigma^2\boldsymbol{\Lambda}_e^{-1}\mathbf{P}^2\boldsymbol{\Lambda}_e^{-1'}), \\ \sqrt{n}(\boldsymbol{\theta}_{o,n} - \boldsymbol{\theta}_o) &\sim N(0, 2\sigma^2\boldsymbol{\Lambda}_e^{-1}\mathbf{P}_\lambda^{2}\boldsymbol{\Lambda}_e^{-1'}), \end{aligned} \tag{8.47C}$$

where

$$\mathbf{P}_\lambda = \text{diag}\left[\frac{\rho_0}{\lambda_0}, \frac{\rho_1}{\lambda_1}, \cdots, \frac{\rho_q}{\lambda_q}\right],$$

and these two $(q + 1)$-vectors become independent in large samples. The formula for the covariances of the odd components results from the fact that in Equation 8.31 we have

$$\boldsymbol{\Lambda}_o = \text{diag}\,[\lambda_0, \lambda_1, \cdots, \lambda_q]\boldsymbol{\Lambda}_e.$$

Hence, in the sampled-continuous case, it is necessary only to invert the $(q + 1)$-by-$(q + 1)$ matrix $\boldsymbol{\Lambda}_e$ to obtain the estimate of the $2(q + 1)$-vector $\boldsymbol{\theta}$.

In the discrete case, we find in precisely the same way that $\sqrt{n}(\boldsymbol{\gamma}_n - \boldsymbol{\gamma})$ has as covariance matrix of its limiting normal distribution, $2\sigma^2\mathbf{Q}^2$, where

$$\mathbf{Q} = \text{diag}\,[\rho_0, \rho_1, \rho_1, \cdots, \rho_q, \rho_q, \rho_{q+1}]$$

is given by Equation 8.46 with the α's and β's now computed from Equation 8.26C rather than Equation 8.26D. Consequently, for the estimate in Step 3, we have

$$\sqrt{n}(\boldsymbol{\theta}_n - \boldsymbol{\theta}) \sim N(0, 2\sigma^2\boldsymbol{\Omega}^{-1}\mathbf{Q}^2\boldsymbol{\Omega}^{-1'}) \tag{8.47D}$$

in $2(q + 1)$ dimensions. The entries in **P** and **Q**, respectively, are given by the formulas

$$\rho_k = \left[\sum_{j=0}^{q} \theta_{2j}(-1)^j \lambda_k^{2j}\right]^2 + \left[\sum_{j=0}^{q} \theta_{2j+1}(-1)^j \lambda_k^{2j+1}\right]^2 \qquad (k = 0, 1, \cdots, q), \quad (8.48\text{C})$$

and

$$\rho_k = \left|\sum_{j=0}^{2q+1} \theta_j e^{ij\omega_k}\right|^2 \qquad (k = 0, 1, \cdots, q + 1). \qquad (8.48\text{D})$$

The limiting distributions 8.47C and 8.47D depend on the unknown θ's only via the values of the ρ's. For the latter situation, we add to Step 2 the calculation of

$$\rho_{k,n} = \alpha_{k,n}^2 + \beta_{k,n}^2,$$

and let $\mathbf{Q}_n$ denote the matrix **Q** with ρ_k replaced by its consistent estimate $\rho_{k,n}$. Then we have

$$\mathbf{Q}_n^{-1}\Omega\sqrt{n}(\boldsymbol{\theta}_n - \boldsymbol{\theta}) \sim N(0, 2\sigma^2\mathbf{I}).$$

If the noise variance σ^2 is unknown, it can be consistently estimated by adding to Step 1 the calculation of

$$s_n^2 = \frac{1}{n - 1 - p} \sum_{t=1}^{n} (\mathbf{Y}_t - \mathbf{h}_t'\boldsymbol{\xi}_n)^2.$$

A similar procedure can be carried out in the sampled-continuous case. Consequently, we can set up *large-sample confidence regions on* $\boldsymbol{\theta}$.

When the independent errors share a common normal distribution, $\boldsymbol{\theta}_n$ *is the Maximum-Likelihood estimate of* $\boldsymbol{\theta}$ for every $n > p + 1$ in both the sampled-continuous and discrete-time cases. This is true because the least-squares estimate $\boldsymbol{\xi}_n$ of $\boldsymbol{\xi}$ becomes the Maximum-Likelihood estimate, and the Maximum-Likelihood estimate of the 1-1 vector-valued function which relates $\boldsymbol{\theta}$ to $\boldsymbol{\xi}$ is the function of the Maximum-Likelihood estimate, namely, $\boldsymbol{\theta}_n$.

This optimum property is conditional on the given regression vectors $\mathbf{h}_1, \mathbf{h}_2, \cdots$. There remains the problem of delineating them by an appropriate *choice of the input frequencies*. In the discrete-time case, the answer, at least from the computational point of view, is clear. The particular selection

$$\omega_k = \frac{\pi k}{q + 1} \qquad (k = 1, 2, \cdots, q) \qquad (8.49)$$

makes Equation 8.35, after normalization, an orthogonal matrix and thereby obviates inversion in Step 3. Using Equation 8.41, we find that

$$\boldsymbol{\Omega}\boldsymbol{\Omega}' = \text{diag}\,[2(q+1), q+1, \cdots, q+1, 2(q+1)].$$

Consequently,

$$\theta_j = \frac{\alpha_0}{2(q+1)} + \frac{1}{q+1}\sum_{k=1}^{q} (\alpha_k \cos j\omega_k + \beta_k \sin j\omega_k) + \frac{(-1)^j}{(2q+1)}\alpha_{q+1}$$

$$(j = 0, 1, \cdots, 2q+1), \qquad (8.50)$$

and the estimate of θ_j is obtained by merely substituting for α_k and β_k the quantities $\alpha_{k,n}$ and $\beta_{k,n}$ which result from Step 2. The limiting covariance matrix in 8.47D reduces to a Toeplitz matrix with entries

$$\frac{\sigma^2}{2(q+1)^2}\left[\rho_0{}^2 + 4\sum_{k=1}^{q} \rho_k{}^2 \cos(j-j')\omega_k + \rho_{q+1}^2(-1)^{j-j'}\right]$$

$$(j, j' = 0, 1, \cdots, 2q+1),$$

where the ρ's are given by 8.48D and 8.49.

In the sampled-continuous case, the choice of $0 < \lambda_0 < \lambda_1 < \cdots < \lambda_q$ is not so obvious. To carry out Step 3, we must invert the $(q+1)$-by-$(q+1)$ matrix $\boldsymbol{\Lambda}_e$, given in (8.31*b*). A procedure for doing this is given in Lemma 8. (Take $r_j = -\lambda_{j-1}^2$ and $n = q+1$; then the conclusion gives the row vectors of the inverse of $\boldsymbol{\Lambda}_e'$.) An analysis of the method would show that certain choices for the λ's make the inversion numerically difficult.

On the other hand, we would like to pick these input frequencies to make our estimate statistically accurate, which we measure by the determinant of the limiting covariance (called the generalized variance). In this regard, it is unimportant how we label the parameters; therefore, the determinant of the limiting covariance matrix of $\sqrt{n}(\boldsymbol{\theta}_n - \boldsymbol{\theta})$ is simply the product of the determinants of the two matrices in 8.47C. The square root of this generalized variance is proportional to

$$G(\lambda_0, \lambda_1, \cdots, \lambda_q; \boldsymbol{\theta}) = \frac{|\mathbf{P}|^2}{|\boldsymbol{\Lambda}_e|^2 \prod_{k=0}^{q} \lambda_k}$$

$$= \frac{|\mathbf{P}|^2}{\prod_{j<k} (\lambda_j{}^2 - \lambda_k{}^2)^2 \prod_{k=0}^{q} \lambda_k}. \qquad (8.51)$$

The numerator is $\prod_{k=0}^{q} \rho_k{}^2$, where ρ_k is given by 8.48C and depends on the unknown parameters. For given bounds on the components of $\boldsymbol{\theta}$, the function in Equation 8.51 can be examined for any particular choice of the input frequencies.

8.5 Elliptical Trajectory Parameter Estimation

Example 8.5. To a first approximation, the trajectory of a small Earth satellite is an ellipse with one of its focii located at the Earth's center of mass. If a polar-coordinate system is chosen in the plane of this ellipse (the origin being located at the Earth's center of mass), the (r, Ψ)-coordinates of the satellite at any time t satisfy the equation

$$r(t) = \frac{a(1 - e)^2}{1 + e \cos(\Psi(t) - \alpha)}, \tag{8.52}$$

where $r(t)$ is the distance from the Earth's center of mass to the satellite at time t, $\Psi(t)$ is the angle between a radius vector from the Earth's center of mass to the satellite and the reference direction of the coordinate system, a is the length of the ellipse's major semiaxis, e is the eccentricity of the ellipse, and α is the angle between the ellipse's major axis and the reference direction. (See Figure 8.3.)

Noisy observations $y_1(t)$, $y_2(t)$, and $y_3(t)$ are made on $r(t)$, $\Psi(t)$, and $\dot{r}(t) = dr/dt$, respectively. Thus we have

$$y_1(t) = r(t) + z_1(t), \qquad y_2(t) = \Psi(t) + z_2(t), \qquad y_3(t) = \dot{r}(t) + z_3(t). \tag{8.53}$$

We wish to reconstruct $r(t)$ and $\Psi(t)$ from the noisy data, so that the position of the satellite can be predicted at any instant of time. We begin our analysis by deriving parametric representations of r and Ψ. The functional forms of $r(t)$ and $\Psi(t)$, which depend upon the parameters a, e, α, and $\Psi(0)$, can be deduced from Newton's laws.

In polar coordinates, the "$F = ma$" equations become

$$a_r = \ddot{r} - r\dot{\Psi}^2 = -\mu/r^2 \qquad (\mu \text{ a known constant}), \tag{8.54}$$

$$a_\Psi = r\ddot{\Psi} + 2\dot{r}\dot{\Psi} = 0, \tag{8.55}$$

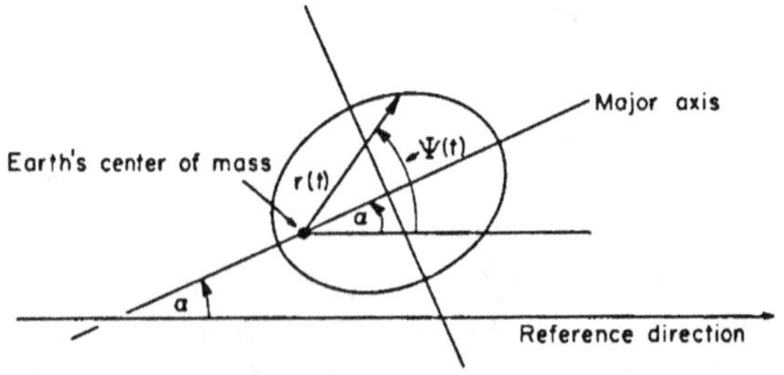

Figure 8.3 Elliptical trajectory of a small Earth satellite.

where the dots denote time derivatives throughout the example. Equation 8.55 can be rewritten

$$\frac{1}{r}\frac{d}{dt}(r^2\dot{\Psi}) = 0.$$

The last equation implies that $r^2\dot{\Psi} =$ const. Thus, we have

$$r^2\dot{\Psi} = M, \tag{8.56}$$

which expresses the conservation of angular momentum.

Here M is related to a and e. If Equation 8.52 is differentiated with respect to t and M/r^2 is substituted for $\dot{\Psi}$, there results

$$\dot{r}(t) = \frac{Me \sin(\Psi(t) - \alpha)}{a(1 - e^2)}. \tag{8.57}$$

Now, we differentiate $\dot{r}$ and use the same substitution:

$$\ddot{r}(t) = \frac{M^2 e \cos(\Psi(t) - \alpha)}{a(1 - e^2)r^2}. \tag{8.58}$$

We substitute 8.56 through 8.58 into Equation 8.54. Thus we obtain

$$\frac{M^2 e \cos(\Psi - \alpha)}{a(1 - e^2)r^2} - \frac{M^2}{r^3} = -\frac{\mu}{r^2}. \tag{8.59}$$

Finally, we substitute 8.52 into Equation 8.59 and solve for M^2. We find that

$$M^2 = a\mu(1 - e^2); \tag{8.60}$$

therefore, Equation 8.56 becomes

$$r^2\dot{\Psi} = (a\mu(1 - e^2))^{1/2}. \tag{8.61}$$

Substituting 8.52 into Equation 8.61, we can integrate the differential equation:

$$\int_{\Psi(0)-\alpha}^{\Psi(t)-\alpha} (a(1 - e^2))^{3/2}\, \mu^{-1/2} \frac{d\Psi}{(1 + e\cos\Psi)^2} = t. \tag{8.62}$$

Equation 8.62 expresses $\Psi(t)$ as an implicit function of four parameters ($\Psi(0)$, α, e, and a).

If $\Psi(t)$ could be solved for explicitly, the resulting expression could be substituted into Equation 8.52, thereby causing $r(t)$ [hence $\dot{r}(t)$] to be represented as functions of these parameters. Unfortunately, the integral 8.62 cannot be represented in terms of elementary functions. We must consequently resort to a clever change of variable.

Before proceeding, let us point out that we have greatly simplified the problem by assuming that the plane of the orbit is known exactly, thereby reducing the number of unknown parameters by two. We will now add one more simplifying assumption, namely, that a (the length of the major semiaxis) is known. Under this assumption, we can choose the unit of length so that

$$a = 1.$$

Since μ has the dimensionality of cubed length over squared time, we can also choose the unit of time so that

$$\mu = 1.$$

Fundamental Equations 8.52 and 8.61 become

$$r(t) = (1 - e^2)/\{1 + e\cos(\Psi(t) - \alpha)\}, \tag{8.63}$$

$$r^2(t)\dot{\Psi}(t) = \sqrt{1 - e^2}. \tag{8.64}$$

Now, we consider the following change of variable:

$$E(t) = \left[\frac{\Psi(t) - \alpha}{2\pi}\right] 2\pi \begin{cases} \arccos\left(\dfrac{e + \cos(\Psi(t) - \alpha)}{1 + e\cos(\Psi(t) - \alpha)}\right) & \text{if } \sin(\Psi - \alpha) \geq 0, \\ 2\pi - \arccos\left(\dfrac{e + \cos(\Psi(t) - \alpha)}{1 + e\cos(\Psi(t) - \alpha)}\right) & \text{if } \sin(\Psi - \alpha) < 0. \end{cases} \tag{8.65}$$

(Here, $[(\Psi - \alpha)/2\pi]$ is the greatest integer in $(\Psi - \alpha)/2\pi$.) As $(\Psi - \alpha)$ varies from 0 to ∞, so does E (and in a monotone fashion). Furthermore, if $k\pi \leq \Psi - \alpha \leq (k + 1)\pi$, the same holds for E: $k\pi \leq E \leq (k + 1)\pi$. In fact,

$$E(t) = k\pi \quad \text{whenever} \quad \Psi(t) - \alpha = k\pi \quad (k = 1, 2, \cdots).$$

As an immediate consequence, the transformation can be inverted as follows:

$$\Psi(t) - \alpha = \left[\frac{E(t)}{2\pi}\right] 2\pi \begin{cases} \arccos\left(\dfrac{e - \cos E(t)}{e\cos E(t) - 1}\right) & \text{if } \sin E \geq 0, \\ 2\pi - \arccos\left(\dfrac{e - \cos E(t)}{e\cos E(t) - 1}\right) & \text{if } \sin E < 0. \end{cases} \tag{8.66}$$

As consequences of Equations 8.65 and 8.66, we obtain

$$\cos E = \frac{e + \cos(\Psi - \alpha)}{1 + e\cos(\Psi - \alpha)}, \qquad 0 \le \Psi - \alpha < \infty, \tag{8.67}$$

and

$$\cos(\Psi - \alpha) = \frac{e - \cos E}{e\cos E - 1}, \qquad 0 \le E < \infty. \tag{8.68}$$

Here $E(t)$ is called the *eccentric anomaly at time t.*

As a consequence of Equations 8.68 and 8.63,

$$r(t) = 1 - e\cos E(t). \tag{8.69}$$

Since $\dot{\Psi}(t) = (d\Psi/dE)(dE/dt)$, we can write Equation 8.64 as

$$r^2(t)\frac{d\Psi}{dE}\frac{dE}{dt} = \sqrt{1 - e^2}. \tag{8.70}$$

Differentiating Equation 8.68 and using 8.69, we find that

$$\sin(\Psi - \alpha)\frac{d\Psi}{dE} = \frac{(1 - e^2)}{r^2}\sin E. \tag{8.71}$$

Computing $\sin(\Psi - \alpha)$ from Equation 8.68, we obtain

$$\sin(\Psi - \alpha) = \frac{(1 - e^2)^{1/2}}{r}\sin E. \tag{8.72}$$

Combining Equations 8.71 and 8.72, we have

$$\frac{d\Psi}{dE} = \frac{(1 - e^2)^{1/2}}{r}. \tag{8.73}$$

After substituting 8.73 into Equation 8.70, we obtain

$$r\frac{dE}{dt} = 1. \tag{8.74}$$

Now, we use 8.69 in Equation 8.74 and integrate:

$$\int_{E(0)}^{E(t)} (1 - e\cos E)\, dE = t.$$

This yields

$$E(t) - e\sin E(t) = t + (E(0) - e\sin E(0)). \tag{8.75}$$

The quantity $E(t) - e\sin E(t)$ is called the *mean anomaly at time t.*

We will parametrize the unknowns as follows:

$$\theta_1 = e, \qquad \theta_2 = E(0) - e\sin E(0), \qquad \text{and} \qquad \theta_3 = \alpha. \tag{8.76}$$

We have chosen θ_2 to be the mean anomaly at time zero instead of $\Psi(0)$, because this parametrization is more useful in orbit determination. It enters explicitly in the representation for $E(t)$:

$$E(t) - \theta_1 \sin E(t) = t + \theta_2. \tag{8.77}$$

Since $x - \theta_1 \sin x$ is monotone increasing when $0 < \theta_1 < 1$ (which it must be for an ellipse), we can solve Equation 8.77 for $E(t)$ as a function of θ_1 and θ_2. Letting $\boldsymbol{\theta}$ be the column vector whose components are defined in Equation 8.76, we can write Equations 8.69 and 8.66 as

$$r(t; \boldsymbol{\theta}) = 1 - \theta_1 \cos E(t), \tag{8.78}$$

$$\Psi(t; \boldsymbol{\theta}) = \theta_3 + \left[\frac{E(t)}{2\pi}\right] 2\pi + \begin{cases} \arccos\left(\dfrac{\theta_1 - \cos E(t)}{\theta_1 \cos E(t) - 1}\right) & \text{if } \sin E \geq 0, \\ 2\pi - \arccos\left(\dfrac{\theta_1 - \cos E(t)}{\theta_1 \cos E(t) - 1}\right) & \text{if } \sin E < 0, \end{cases} \tag{8.79}$$

the dependence of E on the parameters being suppressed in Equation 8.79 to save space. In the sequel, we will generally express E's dependence upon θ_1 and θ_2 by writing $E(t; \boldsymbol{\theta})$ instead of $E(t; \theta_1, \theta_2)$, it being understood that E's dependence on θ_1 and θ_2 is given implicitly by Equation 8.77.

We are now able to set up the desired recursive-estimation scheme. We will assume that bounds are known for θ_1, θ_2, and θ_3:

$$0 < e_1 \leq \theta_1 \leq e_2 < 1, \quad 0 \leq \varphi_1 \leq \theta_2 \leq \varphi_2, \quad \alpha_1 \leq \theta_3 \leq \alpha_2. \tag{8.80}$$

Hence, the truncation set will be a rectangular parallelepiped:

$$\mathscr{P} = [e_1, e_2] \otimes [\varphi_1, \varphi_2] \otimes [\alpha_1, \alpha_2].$$

We will estimate $\boldsymbol{\theta}$ via a truncated, batch-processing, "quick and dirty," adaptive recursion: Let τ be a sampling interval chosen so that the residual vectors

$$\mathbf{Z}_n = \begin{bmatrix} z_1(n\tau) \\ z_2(n\tau) \\ z_3(n\tau) \end{bmatrix}$$

are independent. Also let

$$t_n = n\tau \qquad (n = 1, 2, \cdots),$$

and

$$\mathbf{Y}_n = \begin{bmatrix} y_1(t_n) \\ y_2(t_n) \\ y_3(t_n) \end{bmatrix}, \quad \text{and} \quad \mathbf{F}_n(\boldsymbol{\theta}) = \begin{bmatrix} r(t_n; \boldsymbol{\theta}) \\ \Psi(t_n; \boldsymbol{\theta}) \\ \dot{r}(t_n; \boldsymbol{\theta}) \end{bmatrix}$$

The y's and z's are defined in 8.53. Let $\mathbf{s}_1 = \boldsymbol{\theta}_0 \in \mathscr{P}$, and

$$\mathbf{s}_{n+1} = [\mathbf{s}_n + \mathbf{A}_n(\mathbf{Y}_n - \mathbf{F}_n(\mathbf{s}_n))]_{\mathscr{P}} \qquad (n = 1, 2, \cdots),$$

where $\mathbf{A}_n$ is a 3×3 matrix:

$$\mathbf{A}_n = \left(\sum_{j=1}^{3n} \|\mathbf{h}_j\|^2\right)^{-1} (\mathbf{h}_{3n-2}, \mathbf{h}_{3n-1}, \mathbf{h}_{3n}),$$

and

$$\left.\begin{aligned} \mathbf{h}_{3n-2} &= \operatorname{grad} r(t_n; \boldsymbol{\theta})|_{\boldsymbol{\theta}=\mathbf{s}_n} \\ \mathbf{h}_{3n-1} &= \operatorname{grad} \Psi(t_n; \boldsymbol{\theta})|_{\boldsymbol{\theta}=\mathbf{s}_n} \\ \mathbf{h}_{3n} &= \operatorname{grad} \dot{r}(t_n; \boldsymbol{\theta})|_{\boldsymbol{\theta}=\mathbf{s}_n} \end{aligned}\right\} \qquad (n = 1, 2, \cdots). \tag{8.81}$$

The following formulas are necessary to carry out the recursion: Combining Equation 8.57 with Equation 8.60 (and remembering that $a = \mu = 1$ in this example), we find that

$$\dot{r}(t; \boldsymbol{\theta}) = \theta_1(1 - \theta_1^2)^{-1/2} \sin(\Psi(t; \boldsymbol{\theta}) - \theta_3), \tag{8.82}$$

which, together with Equations 8.78 and 8.79, define the components of $\mathbf{F}_n$. Furthermore, we have

$$\operatorname{grad} r(t; \boldsymbol{\theta}) = \begin{bmatrix} -\cos(\Psi(t; \boldsymbol{\theta}) - \theta_3) \\ \theta_1(1 - \theta_1^2)^{1/2} \sin(\Psi(t; \boldsymbol{\theta}) - \theta_3) \\ 0 \end{bmatrix} \tag{8.83}$$

$$\operatorname{grad} \Psi(t; \boldsymbol{\theta}) = \begin{bmatrix} \left(\dfrac{1}{r(t; \boldsymbol{\theta})} - \dfrac{1}{1 - \theta_1^2}\right) \sin(\Psi(t; \boldsymbol{\theta}) - \theta_3) \\ (1 - \theta_1^2)^{1/2}/r^2(t; \boldsymbol{\theta}) \\ 1 \end{bmatrix}, \tag{8.84}$$

and

$$\operatorname{grad} \dot{r}(t; \boldsymbol{\theta}) = \begin{bmatrix} (1 - \theta_1^2)^{1/2} \sin(\Psi(t; \boldsymbol{\theta}) - \theta_3)/r^2(t; \boldsymbol{\theta}) \\ \theta_1 \cos(\Psi(t; \boldsymbol{\theta}) - \theta_3)/r^2(t; \boldsymbol{\theta}) \\ 0 \end{bmatrix}. \tag{8.85}$$

(Equations 8.83 through 8.85 will be derived at the end of this example.)

A typical computation cycle might go like this, where θ_{1n}, θ_{2n}, and θ_{3n} denote the components of the (column) vector $\mathbf{s}_n$.

1. Substitute $\mathbf{s}_n$ for $\boldsymbol{\theta}$ in Equation 8.77 and solve for $E(t_n; \mathbf{s}_n)$.
2. Compute $r(t_n; \mathbf{s}_n)$ from Equation 8.78.
3. Compute $\Psi(t_n; \mathbf{s}_n)$ from Equation 8.79.
4. Compute $\cos[\Psi(t_n; \mathbf{s}_n) - \theta_{3n}]$, and $\sin[\Psi(t_n; \mathbf{s}_n) - \theta_{3n}]$ from Equation 8.68.
5. Compute $\dot{r}(t_n; \mathbf{s}_n)$ from Equation 8.82.
6. Compute $\mathbf{h}_{3n-2}, \mathbf{h}_{3n-1}, \mathbf{h}_{3n}$, using Equations 8.81 and 8.83 through 8.85.
7. Update $\sum_{j=1}^{3n-3} \|\mathbf{h}_j\|^2$ to $\sum_{j=1}^{3n} \|\mathbf{h}_j\|^2$ and form $\mathbf{A}_n$.
8. Form the column vector $\mathbf{F}_n(\mathbf{s}_n)$ of quantities in (2), (3), and (5).
9. Observe $\mathbf{Y}_n$ and compute $[\mathbf{s}_n + \mathbf{A}_n(\mathbf{Y}_n - \mathbf{F}_n(\mathbf{s}_n))]_{\mathscr{P}} = \mathbf{s}_{n+1}$.
10. Begin the next cycle.

The gains that we use are given by Equation 7.48*c*, so we *verify Assumptions F1 through F5 of Theorem 7.1.* We begin by pointing out the following:

$$0 < \inf_{\boldsymbol{\theta}\in\mathscr{P}, t>0} \begin{Bmatrix} \|\text{grad } r(t; \boldsymbol{\theta})\| \\ \|\text{grad } \Psi(t; \boldsymbol{\theta})\| \\ \|\text{grad } \dot{r}(t; \boldsymbol{\theta})\| \end{Bmatrix} \leq \sup_{\boldsymbol{\theta}\in\mathscr{P}, t>0} \begin{Bmatrix} \|\text{grad } r(t; \boldsymbol{\theta})\| \\ \|\text{grad } \Psi(t; \boldsymbol{\theta})\| \\ \|\text{grad } \dot{r}(t; \boldsymbol{\theta})\| \end{Bmatrix} < \infty. \tag{8.86}$$

This follows from Equations 8.83 through 8.85 by virtue of Equation 8.80.

Straightforward differentiation of Equations 8.83 through 8.85 will verify that each element of the matrices of second-order mixed partial ($\boldsymbol{\theta}$) derivatives of r, Ψ, and $\dot{r}$ is uniformly bounded for $\boldsymbol{\theta} \in \mathscr{P}$ and $t > 0$. Thus,

$$\limsup_n \left[\sup_{\mathbf{x}_1\in\mathscr{P}, \mathbf{x}_2\in\mathscr{P}, \mathbf{x}_3\in\mathscr{P}} \|\mathbf{G}_n(\mathbf{x}_1, \mathbf{x}_2, \mathbf{x}_3)\|\right] < \infty.$$

If we define $\mathbf{h}_n^*$ as was done in Equation 8.81, except that the derivatives are all evaluated at $\boldsymbol{\theta}_0$ instead of $\mathbf{s}_n$, then, by Equation 8.86, we find that there are constants K_1 and K_2 such that

$$0 < K_1 \leq \|\mathbf{h}_n^*\| \leq K_2 < \infty \tag{8.87}$$

for all n. F1 through F4 now follow immediately.

To prove F5, it suffices to prove that

$$\liminf_{n\to\infty} \lambda_{\min}\left(\sum_{j=1}^{3} \frac{\mathbf{h}_{3n+j}^* \mathbf{h}_{3n+j}^{*\prime}}{\|\mathbf{h}_{3n+j}^*\|^2}\right) > 0.$$

In view of Equation 8.87, it therefore suffices to show that

$$\liminf_{n\to\infty} \lambda_{\min}\left(\sum_{j=1}^{3} \mathbf{h}^*_{3n+j}\mathbf{h}^{*\prime}_{3n+j}\right) = \liminf_{n\to\infty} \lambda_{\min}(\mathbf{H}_n\mathbf{H}_n') > 0, \quad (8.88)$$

where

$$\mathbf{H}_n = (\mathbf{h}^*_{3n+1}, \mathbf{h}^*_{3n+2}, \mathbf{h}^*_{3n+3}).$$

The matrix $\mathbf{H}_n\mathbf{H}_n'$ has three nonnegative eigenvalues

$$0 \le \lambda_{1n} \le \lambda_{2n} \le \lambda_{3n}.$$

Therefore,

$$\lambda_{1n}\lambda_{3n}^2 \ge \lambda_{1n}\lambda_{2n}\lambda_{3n} = \det \mathbf{H}_n\mathbf{H}_n' = (\det \mathbf{H}_n)^2. \quad (8.89)$$

Since

$$\lambda_{3n} \le \lambda_{1n} + \lambda_{2n} + \lambda_{3n} = \operatorname{tr} \mathbf{H}_n\mathbf{H}_n' = \sum_{j=1}^{3} \|\mathbf{h}^*_{3n+j}\|^2, \quad (8.90)$$

it follows from Equations 8.89 and 8.90 that

$$\lambda_{\min}(\mathbf{H}_n\mathbf{H}_n') = \lambda_{1n} \ge \left\{\frac{|\det H_n|}{\sum_{j=1}^{3} \|\mathbf{h}^*_{3n+j}\|^2}\right\}^2$$

In the light of this and Equation 8.87, we find that Equation 8.88 holds if $|\det \mathbf{H}_n|$ is bounded away from zero. If $\det \mathbf{H}_n$ is expanded by cofactors of its last row, we see that

$$|\det \mathbf{H}_n| = \theta_{10}\left\{\frac{\cos^2(\Psi - \theta_{30})}{r^2} + \frac{\sin^2(\Psi - \theta_{30})}{r^2}\right\} = \frac{\theta_{10}}{r^2} \ge \frac{e_1}{r^2}.$$

Since $r(t_n)$ is uniformly bounded, it follows that $\liminf_{n\to\infty} |\det \mathbf{H}_n| > 0$.

We must now derive Equations 8.83 through 8.85. The following identities are basic to the derivations that follow: From Equation 8.67,

$$\cos E + e \cos E \cos(\Psi - \alpha) = e + \cos(\Psi - \alpha),$$

and thus

$$\cos E - e = (1 - e\cos E)\cos(\Psi - \alpha),$$

or

$$\cos E - e = r\cos(\Psi - \alpha), \quad (8.91)$$

after using Equation 8.69 (which we restate for convenience):

$$r = 1 - e\cos E. \quad (8.92)$$

Differentiating Equation 8.92 with respect to time, we obtain

$$\dot{r} = \dot{E} e \sin E.$$

From Equation 8.75,

$$\dot{E}(1 - e\cos E) = 1.$$

and from Equation 8.92,

$$\dot{E}r = 1,$$

and thus

$$\dot{r} = e \sin \frac{E}{r}. \tag{8.93}$$

We differentiate Equation 8.77 with respect to $\theta_1(=e)$:

$$\frac{\partial E}{\partial \theta_1}(1 - \theta_1 \cos E) - \sin E = 0. \tag{8.94}$$

By Equation 8.92,

$$\frac{\partial E}{\partial \theta_1} = \frac{\sin E}{r}. \tag{8.95}$$

We differentiate Equation 8.77 with respect to θ_2:

$$\frac{\partial E}{\partial \theta_2}(1 - \theta_1 \cos E) = 1.$$

Thus, we have

$$\frac{\partial E}{\partial \theta_2} = \frac{1}{r}, \tag{8.96}$$

and finally,

$$\frac{\partial E}{\partial \theta_3} = 0. \tag{8.97}$$

We differentiate Equation 8.92 with respect to θ_1:

$$\frac{\partial r}{\partial \theta_1} = -\cos E + \theta_1 \sin E \frac{\partial E}{\partial \theta_1}.$$

Using Equations 8.92 and 8.95, we obtain

$$\frac{\partial r}{\partial \theta_1} = \frac{\theta_1 - \cos E}{r}. \tag{8.98}$$

By Equation 8.91,

$$\frac{\partial r}{\partial \theta_1} = -\cos(\Psi - \theta_3).$$

Similarly, we have

$$\frac{\partial r}{\partial \theta_2} = \theta_1 \sin E \frac{\partial E}{\partial \theta_2} = \frac{\theta_1 \sin E}{r}. \tag{8.99}$$

By Equation 8.93,

$$\frac{\partial r}{\partial \theta_2} = \theta_1(1 - \theta_1^2)^{1/2} \sin(\Psi - \theta_3), \qquad \text{and} \qquad \frac{\partial r}{\partial \theta_3} = 0.$$

We differentiate Equation 8.68 with respect to θ_1:

$$\frac{\partial \Psi}{\partial \theta_1}(-\sin(\Psi - \theta_3)) = -\frac{1}{r^2}(\cos E - \theta_1)\frac{\partial r}{\partial \theta_1} - \frac{\sin E}{r}\frac{\partial E}{\partial \theta_1} - \frac{1}{r}.$$

We use 8.92, 8.93, 8.95, and 8.98 to show that

$$\frac{\partial \Psi}{\partial \theta_1} = (1 - \theta_1^2)^{-1/2}\frac{\sin E}{r^2}(r + (1 - \theta_1^2)). \tag{8.100}$$

By 8.93, we have

$$\frac{\partial \Psi}{\partial \theta_1} = \left(\frac{1}{r} + \frac{1}{1 - \theta_1^2}\right)\sin(\Psi - \theta_3).$$

We differentiate 8.68 with respect to θ_2:

$$\frac{\partial \Psi}{\partial \theta_2}(-\sin(\Psi - \theta_3)) = \frac{1}{r^2}(\theta_1 - \cos E)\frac{\partial r}{\partial \theta_2} - \frac{1}{r}\sin E\frac{\partial E}{\partial \theta_2}.$$

Using Equations 8.99, 8.96, and 8.93, we obtain

$$\frac{\partial \Psi}{\partial \theta_2} = \frac{(1 - \theta_1^2)^{1/2}}{r^2}, \tag{8.101}$$

and finally

$$\frac{\partial \Psi}{\partial \theta_3} = 1. \tag{8.102}$$

We differentiate 8.93 with respect to θ_1:

$$\begin{aligned}\frac{\partial \dot{r}}{\partial \theta_1} &= \frac{\sin E}{r} + \frac{\theta_1 \cos E}{r}\frac{\partial E}{\partial \theta_1} - \frac{\theta_1 \sin E}{r^2}\frac{\partial r}{\partial \theta_1}\\ &= \frac{\sin E}{r^3}(r^2 + r\theta_1 \cos E - \theta_1^2 + \theta_1 \cos E)\\ &= \left(\frac{1 - \theta_1^2}{r^2}\right)\left(\frac{\sin E}{r}\right)\\ &= \frac{(1 - \theta_1^2)^{1/2}}{r^2}\sin(\Psi - \alpha).\end{aligned}$$

Similarly,

$$\begin{aligned}\frac{\partial \dot{r}}{\partial \theta_2} &= \theta_1\left(\frac{\cos E}{r}\frac{\partial E}{\partial \theta_2} - \frac{\sin E}{r^2}\frac{\partial r}{\partial \theta_2}\right)\\ &= \frac{\theta_1}{r^3}(\cos E - \theta_1) = \frac{\theta_1}{r^2}\cos(\Psi - \alpha).\end{aligned}$$

Finally, we obtain

$$\frac{\partial \dot{r}}{\partial \theta_3} = 0.$$

9. Open Problems

In closing, we wish to call attention to a number of problems that are related to the work contained in this monograph.

9.1 Proof of the Conjectured Theorem

In Chapter 6 we were forced to state as conjecture a theorem pertaining to the almost sure and quadratic-mean convergence of scalar-observation, truncated-estimation recursions of the form

$$\mathbf{t}_{n+1} = [\mathbf{t}_n + \mathbf{a}_n(Y_n - F_n(\mathbf{t}_n))]_{\mathscr{P}},$$

where $\mathscr{P}$ is a closed convex subset of Euclidean p-space.

There is little doubt in our mind regarding the correctness of the theorem, and we hope that one of our readers will have better luck than we did in inventing a correct proof.

9.2 Extensions of Chapters 3 Through 5 to the Vector-Parameter Case

In Chapter 7 we discussed two distinctly different sets of gain sequences: the "linearized least-squares" gains 7.48*b*, *d*, and the "quick and dirty" gains 7.48*a*, *c*. Under a reasonable set of regularity conditions, both types of gains yield convergent estimator sequences. The latter family is unquestionably more convenient from the computational point of view, whereas the former is more efficient in the statistical sense (at least in the case of linear regression). In the general case, it is not unreasonable to expect that a tradeoff exists between computational convenience and statistical efficiency, not only for the classes of gains already discussed but also for any others that one may dream up.

In order to investigate this issue quantitatively, the techniques of Chapters 3 through 5 will have to be extended to the vector-parameter case. Such results will also serve the equally important purpose of allowing one to construct asymptotic confidence sets for nonlinear regression parameters.

9.3 Kalman-Type Filtering Theory for Nonlinear Systems

Throughout this monograph, the parameter that is to be estimated does not change in time. However, in many applications it is desired to estimate a quantity that changes in time, according to an equation of the form

$$\boldsymbol{\theta}_{n+1} = \boldsymbol{\Phi}_n(\boldsymbol{\theta}_n) + \mathbf{V}_n, \tag{9.1}$$

where the functional form of $\boldsymbol{\Phi}_n(\cdot)$ is known, $\mathbf{V}_n$ is a zero-mean stochastic process, and incomplete information about the values of $\boldsymbol{\theta}_n$ is available through observations $\mathbf{Y}_n$ of the form

$$\mathbf{Y}_n = \mathbf{F}_n(\boldsymbol{\theta}_n) + \mathbf{W}_n. \tag{9.2}$$

(When $\boldsymbol{\Phi}_n(\cdot)$ is the identity transformation and $\mathbf{V}_n$ is zero for each n, the problem reduces to ordinary regression.)

When $\mathbf{F}_n(\cdot)$ and $\boldsymbol{\Phi}_n(\cdot)$ are linear functions of their argument and the vector processes $\{\mathbf{V}_n\}$ and $\{\mathbf{W}_n\}$ are mutually and temporally independent, Kalman has developed a recursive theory of smoothing and prediction which generates estimates for $\boldsymbol{\theta}_n$ which are optimal in a number of statistical senses. For example, if $\hat{\boldsymbol{\theta}}_{n|n}$ denotes the estimate of $\boldsymbol{\theta}_n$ based upon the observations $\mathbf{Y}_1, \mathbf{Y}_2, \cdots, \mathbf{Y}_n$, then

$$\hat{\boldsymbol{\theta}}_{n+1|n+1} = \boldsymbol{\Phi}_n(\hat{\boldsymbol{\theta}}_{n|n}) + \mathbf{A}_n[\mathbf{Y}_{n+1} - \mathbf{F}_{n+1}(\hat{\boldsymbol{\theta}}_{n|n})] \tag{9.3}$$

where the smoothing vectors $\mathbf{A}_n$ (or matrices as the case may be) are defined recursively in terms of the second-order noise statistics for $\{\mathbf{V}_n\}$ and $\{\mathbf{W}_n\}$ and the transformations (which, in the linear case, are matrices) $\boldsymbol{\Phi}_n$ and $\mathbf{F}_n$. (See Kalman, 1960.)

Motivated by the computational appeal of the recursive formulation, the prevailing approach in cases when $\boldsymbol{\Phi}_n(\cdot)$ and/or $\mathbf{F}_n(\cdot)$ are nonlinear has been the "method of linearization," coupled with Kalman filtering. Briefly, this approach involves the "linearization" of Equations 9.1 and

9.2, which is accomplished by expanding $\mathbf{\Phi}_n(\cdot)$ and $\mathbf{F}_n(\cdot)$ in a Taylor series about a "nominal value," $\boldsymbol{\theta}_n^\circ$, usually, but not always, chosen to satisfy a noiseless version of the "state equation," Equation 9.1:

$$\boldsymbol{\theta}_{n+1}^\circ = \mathbf{\Phi}_n(\boldsymbol{\theta}_n^\circ). \tag{9.4}$$

When this is done, and if all terms of nonlinear order are ignored, we find that

$$\boldsymbol{\theta}_{n+1} \approx \mathbf{\Phi}_n(\boldsymbol{\theta}_n^\circ) + \dot{\mathbf{\Phi}}_n(\boldsymbol{\theta}_n^\circ)(\boldsymbol{\theta}_n - \boldsymbol{\theta}_n^\circ) + \mathbf{V}_n;$$

therefore, by Equation 9.4,

$$(\boldsymbol{\theta}_{n+1} - \boldsymbol{\theta}_{n+1}^\circ) \approx \mathbf{\Phi}_n(\boldsymbol{\theta}_n^\circ) + \dot{\mathbf{\Phi}}_n(\boldsymbol{\theta}_n^\circ)(\boldsymbol{\theta}_n - \boldsymbol{\theta}_n^\circ) + \mathbf{V}_n, \tag{9.5}$$

and

$$\mathbf{Y}_n - \mathbf{F}_n(\boldsymbol{\theta}_n^\circ) \approx \dot{\mathbf{F}}_n(\boldsymbol{\theta}_n^\circ)(\boldsymbol{\theta}_n - \boldsymbol{\theta}_n^\circ) + \mathbf{W}_n, \tag{9.6}$$

where $\dot{\mathbf{\Phi}}_n(\boldsymbol{\theta}_n^\circ)$ and $\dot{\mathbf{F}}_n(\boldsymbol{\theta}_n^\circ)$ are, respectively, the matrices of $\mathbf{\Phi}_n$ and $\mathbf{F}_n$'s first partial derivatives evaluated at $\boldsymbol{\theta}_n^\circ$.

If the Kalman filtering theory is applied to the linear approximation Equations 9.5 and 9.6, we find that

$$\hat{\boldsymbol{\theta}}_{n+1|n+1} = \mathbf{\Phi}_n(\hat{\boldsymbol{\theta}}_{n|n}) + \mathbf{A}_n[\mathbf{Y}_{n+1} - \mathbf{F}_{n+1}(\hat{\boldsymbol{\theta}}_{n|n})], \tag{9.7}$$

where now $\mathbf{A}_n$ is defined recursively in terms of the second-order noise statistics for $\{\mathbf{W}_n\}$ and $\{\mathbf{V}_n\}$ and in terms of the matrices $\dot{\mathbf{F}}_n(\boldsymbol{\theta}_n^\circ)$ and $\dot{\mathbf{\Phi}}_n(\boldsymbol{\theta}_n^\circ)$.

Although this technique meets with wide acceptance in applications, little if any work (to the best of our knowledge) has been directed toward the analysis of the "steady-state" operating characteristics of such schemes. Of particular interest are such questions as: What is the large-sample (large n) mean-square estimation error of $\hat{\boldsymbol{\theta}}_{n|n}$? What is the quantitative nature of the tradeoff between computational convenience and accuracy that one experiences with various choices of the gains $\mathbf{A}_n$?

The estimation recursion 9.7 looks so much like the recursions for regression-parameter estimation that there is every reason to hope that the analytic approaches developed in this monograph can be carried over and extended to the more general case. Indeed, when the state and observation Equations 9.1 and 9.2 are scalar relations, our previous methods can be applied and furnish a bound on the limiting mean-square prediction error.

From the first n observations, we recursively predict θ_{n+1} by $t_{n+1}(=\theta_{n+1|n}$ in the previous notation):

$$t_{n+1} = \Phi_n(t_n) + a_n[Y_n - F_n(t_n)] \qquad (n = 1, 2, \cdots). \tag{9.8}$$

Here t_1 is an arbitrary random variable estimating the initial state θ_1 (each assumed to have finite second-order moments). We use gains $\{a_n\}$ that minimize, subject to a certain constraint, the steady-state prediction error under the following assumptions:

H1. The noise processes $\{V_n\}$ and $\{W_n\}$ are mutually and temporally independent with zero means and some finite (nonzero) variances σ_V^2 and σ_W^2.

H2. The derivatives $\dot{\Phi}_n(x)$ and $\dot{F}_n(x)$ are one-signed for each n.

H3. For all x, $\beta_n' \leq |\dot{\Phi}_n(x)| \leq \beta_n$ and $b_n \leq |\dot{F}_n(x)| \leq b_n'$, where $\beta_n \leq c_1\beta_n'$ and $b_n' \leq c_2 b_n$ for some $1 \leq c_1, c_2 < \infty$.

H4. $\beta = \limsup_n \beta_n < \dfrac{c}{c-1}$, where $c = c_1 c_2$.

H5. $b_n \to \infty$ as $n \to \infty$.

The significance of the last two assumptions will be discussed after we prove the following theorem.

THEOREM 9.1

Let $\{\theta_n\}$ and $\{Y_n\}$ be scalar-valued processes defined by Equations 9.1 and 9.2 which satisfy Assumptions H1 through H5. Let $\{t_n\}$ be generated by Equation 9.8 with

$$a_n = \operatorname{sgn}(\dot{\Phi}_n \dot{F}_n)\frac{\beta_n}{cb_n}.$$

Then, for the mean-square prediction error, we have

$$\limsup_n \mathscr{E}(t_n - \theta_n)^2 \leq \frac{\sigma_V^2}{1 - \beta^2\left(1 - \dfrac{1}{c}\right)^2}$$

with equality in the linear case (when $c = 1$).

Proof. The usual Taylor-series expansion gives

$$t_{n+1} - \theta_{n+1} = (\dot{\Phi}_n - a_n\dot{F}_n)(t_n - \theta_n) - V_n + a_n W_n,$$

where the derivatives are evaluated at some intermediate points. We square this, take expectations conditional on $\theta_1, V_1, \cdots, V_{n-1}$; $t_1, W_1, \cdots, W_{n-1}$, and use Assumption H1. The result combines with Assumptions H2 and H3 to yield

$$e_{n+1}^2 \leq (\beta_n - |a_n|b_n)^2 e_n^2 + \sigma_V^2 + \sigma_W^2 a_n^2, \tag{9.9}$$

where $e_n^2 = \mathscr{E}(t_n - \theta_n)^2$. A lower inequality holds with $\beta_n - |a_n|b_n$ replaced by $\beta_n' - |a_n|b_n' \geq 0$. Let us set

$$B_j = (\beta_j - |a_j|b_j)^2, \qquad B_{nk} = (1 - B_k)\prod_{j=k+1}^{n} B_j. \tag{9.10}$$

Then, after iterating Equation 9.9 back to (say) $n = 1$, we have

$$e_{n+1}^2 \leq \prod_{j=1}^{n} B_j \cdot e_1^2 + \sum_{k=1}^{n} B_{nk} \frac{\sigma_V^2 + \sigma_W^2 a_k^2}{1 - (\beta_n - |a_k| b_k)^2}. \tag{9.11}$$

By Assumption H4 and the choice of $|a_j|$, we find that

$$\beta_j - |a_j| b_j = \beta_j \left(1 - \frac{1}{c}\right) \tag{9.12}$$

is bounded away from unity for all large enough j, so the leading product in Equation 9.11 goes to zero as n tends to infinity. According to Lemma 1,

$$\sum_{k=1}^{n} B_{nk} \equiv 1 - \prod_{j=1}^{n} B_j \to 1.$$

It remains to apply Equation 4.20*a* to the summation in Equation 9.11 to conclude that

$$\limsup_n e_n^2 \leq \limsup_n \frac{\sigma_V^2 + \sigma_W^2 a_n^2}{1 - (\beta_n - |a_n|\, b_n)^2} = \frac{\sigma_V^2}{1 - \beta^2(1 - (1/c))^2}. \tag{9.13}$$

The equality is a consequence of Equation 9.12, Assumption H4, and Assumption H5. Q.E.D.

It is important to note that Assumption H4 does not require "stability" of the state equation $\theta_{n+1} = \Phi_n(\theta_n) + V_n$, except in the limit of indefinitely large c_1 or c_2. At the other end of the spectrum when $\beta_n' = \beta_n = \beta$ and $b_n = b_n'$ (the linear case), Assumption H4 allows systems $\theta_{n+1} = \pm\beta\theta_n + V_n$ with $\beta < \infty$ arbitrary. The reason for this, of course, is Assumption H5 which states that $\inf_x |\dot{F}_n(x)| \to \infty$. A smaller and smaller error therefore results in estimating θ_n, whose fluctuations are increasing without bound from an observation Y_n. In such cases, the result is exact:

$$\lim_n e_n^2 = \sigma_V^2.$$

This is the same error that would obtain at each step if we were able to observe the state variable directly.

By stating our result in a slightly more general form, we can see why we selected the particular gains in the statement of the theorem and also investigate the situation when $b_n \to 0$. We define, for positive numbers x, the sequence of functions

$$Q_n(x) = \frac{\sigma_V^2 + \sigma_W^2 x^2}{1 - (\beta_n - x b_n)^2}. \tag{9.14}$$

By filling in the details of the proof, it is not difficult to see that under Assumptions H1 through H3 the first line in Equation 9.13 holds true, namely,

$$\limsup_n e_n^2 \leq \limsup_n Q_n(|a_n|), \tag{9.15}$$

provided that

$$|a_n| \leq \frac{\beta_n'}{b_n'}, \qquad \operatorname{sgn} a_n = \operatorname{sgn}(\dot{\Phi}_n \dot{F}_n), \tag{9.16a}$$

and

$$\prod_{j=1}^{n} (\beta_j - |a_j|\, b_j)^2 \to 0 \tag{9.16b}$$

as $n \to \infty$. For given sequences $\{\beta_n\}$ and $\{b_n\}$, the function 9.14 has a unique minimum at the point

$$x_0 = \frac{-R_n + \sqrt{R_n^2 + 4\sigma_V^2\sigma_W^2}}{2\sigma_W^2} \tag{9.17}$$

where

$$R_n = \frac{\sigma_W^2(1 - \beta_n^2) + \sigma_V^2 b_n^2}{\beta_n b_n}. \tag{9.18}$$

Under Assumption H5, when $b_n \to \infty$, we have

$$R_n \cong \frac{\sigma_V^2 b_n}{\beta_n} \to \infty$$

as $n \to \infty$, assuming always that β_n has a finite limit superior. Consequently,

$$\sqrt{R_n^2 + 4\sigma_V^2\sigma_W^2} = R_n + \frac{2\sigma_V^2\sigma_W^2}{R_n} + o\left(\frac{1}{R_n}\right)$$

and

$$x_0 \cong \frac{\beta_n}{b_n} \to 0 \tag{9.19}$$

independent of the unknown variances. If we were to use this x_0 as $|a_n|$, we would not meet the condition of Equation 9.16*a*; hence the reason for the division by c. Assumption H4 ensures that Equation 9.16*b* is satisfied.

As one might expect, the situation is entirely different when

H5′. $b_n \to 0$.

In this case, we have for Equation 9.18

$$R_n \cong \frac{\sigma_W^2(1 - \beta_n^2)}{\beta_n b_n} \to \infty,$$

assuming that β_n does not tend to unity. In the same way as in the previous paragraph, we find for Equation 9.17 that

$$x_0 \cong \frac{\sigma_V^2}{\sigma_W^2} \frac{\beta_n}{1 - \beta_n^2} b_n \to 0$$

as n tends to infinity. Since σ_V^2/σ_W^2 is not known, this suggests using gains with

$$|a_n| = a \frac{\beta_n}{1 - \beta_n^2} b_n \qquad (a > 0), \tag{9.20}$$

and assuming, in place of Assumption H4, that

H4′. $\beta < 1$.

Then Equation 9.16a holds, at least for all large enough n (which is enough), as does Equation 9.16b. Thus, for the gains $a_n = \operatorname{sgn}(\dot{\Phi}_n \dot{F}_n)\,|a_n|$ in Equation 9.20, the result of Equation 9.15 reads

$$\limsup_n e_n^2 \le \frac{\sigma_V^2}{1 - \beta^2}, \tag{9.21}$$

and there is equality when $\beta_n' = \beta_n = \beta$ (and lim sup should be replaced by lim). But this is precisely the mean-square error resulting from not using the observations at all (that is, by setting $a = 0$ in Equation 9.20). In other words, for a stable system

$$\theta_{n+1} = \pm\beta\theta_n + V_n \qquad (0 < \beta < 1)$$

and an observational equation

$$Y_n = F_n(\theta_n) + W_n \quad \text{with} \quad \sup_x |\dot{F}_n(x)| \le c_2 \inf_x |\dot{F}_n(x)| \to 0,$$

the "observation-free" predictor $t_{n+1} = \pm\beta t_n$ does just as well as the "optimized" version of Equation 9.8.

Finally, when $b_n = b_n'$ tends to a nonzero finite constant (say unity) and $\beta_n' = \beta_n = \beta$ $(0 < \beta < 1)$, the same approach will lead to time-independent gains $|a_n| = a$, where

$$a = \beta \frac{Q_0}{Q_0 + \sigma_W^2} \tag{9.22}$$

and

$$Q_0 = \sigma_V^2 + a\beta\sigma_W^2. \tag{9.23}$$

These two equations combine to give a quadratic (Kalman's "variance equation") whose positive square-root solution is the minimum mean-square linear prediction error Q_0. The optimum gain now depends on σ_V^2 and σ_W^2 as well as β. This result is a special case of Kalman's linear theory (see his Example 1), and we include it only as a point of comparison.

Appendix. Lemmas 1 Through 8

Lemma 1

Let $\mathbf{A}_1, \mathbf{A}_2, \cdots$ be a sequence of square matrices. Then, for all $1 \leq k \leq n$ and $n \geq 1$,

$$\sum_{j=k}^{n} \prod_{i=j+1}^{n} (\mathbf{I} - \mathbf{A}_i)\mathbf{A}_j = \mathbf{I} - \prod_{i=k}^{n} (\mathbf{I} - \mathbf{A}_i),$$

where products are to be read backwards and void ones defined as the identity.

Proof. We have

$$\begin{aligned}\prod_{i=j+1}^{n} (\mathbf{I} - \mathbf{A}_i) - \prod_{i=j}^{n} (\mathbf{I} - \mathbf{A}_i) &= \prod_{i=j+1}^{n} (\mathbf{I} - \mathbf{A}_i)[\mathbf{I} - (\mathbf{I} - \mathbf{A}_j)] \\ &= \prod_{i=j+1}^{n} (\mathbf{I} - \mathbf{A}_i)\mathbf{A}_j.\end{aligned}$$

Thus,

$$\begin{aligned}\prod_{i=j+1}^{n} (\mathbf{I} - \mathbf{A}_i)\mathbf{A}_j = &\left[\prod_{i=j+1}^{n} (\mathbf{I} - \mathbf{A}_i) - \prod_{i=k}^{n} (\mathbf{I} - \mathbf{A}_i)\right] \\ &- \left[\prod_{i=j}^{n} (\mathbf{I} - \mathbf{A}_i) - \prod_{i=k}^{n} (\mathbf{I} - \mathbf{A}_i)\right].\end{aligned}$$

The sum over j from k to n of the right-hand side collapses to yield the asserted result. Q.E.D.

Lemma 2

Let $P_n = \prod_{j=1}^{n} (1 - a_j)$ when $a_j \in (0, 1)$ for all $j \geq N$ and $P_n \to 0$ as $n \to \infty$. Then, if $\sum_k x_k < \infty$,

$$\max_{1 \leq k \leq n} \sum_{j=k}^{n} \frac{P_n}{P_j} x_j \to 0$$

as $n \to \infty$.

Proof. The maximum in question is equal to the larger of the two values which result from maximizing over $1 \leq k \leq N - 1$ and over $N \leq k \leq$ n. The former is $O(P_n) = o(1)$ as $n \to \infty$, and we must prove the latter is also. To save writing, we set $R_{j+1} = P_n/P_j$ and $s_{j+1} = x_1 + \cdots + x_j$ (where all void products are to be read as unity). Then we have the identity

$$\begin{aligned}\sum_{j=k}^{n} R_{j+1} x_j &= \sum_{j=k}^{n} (R_{j+1} s_{j+1} - R_{j+1} s_j) \\ &= \sum_{j=k}^{n} (R_j s_j - R_{j+1} s_j) + s_{n+1} - R_k s_k \\ &= -\sum_{j=k}^{n} R_{j+1} a_j s_j + s_{n+1} - R_k s_k.\end{aligned}$$

But, according to Lemma 1, the last expression is unaltered if we subtract a constant from every subscripted s. Using $s = \lim_n s_n$, we therefore have

$$\left|\sum_{j=k}^{n} \frac{P_n}{P_j} x_j\right| \leq |s_{n+1} - s| + \left|\frac{P_n}{P_{k-1}} (s_k - s)\right| + \left|\sum_{j=k}^{n} \frac{a_j P_n}{P_j} (s_j - s)\right| \tag{L2.1}$$

in the original notation. With regard to the second term, given $\varepsilon > 0$ we choose $n_1 > N$ so that $|s_k - s| < \varepsilon$ for all $k > n_1$, and then $n_2 > n_1$ so that $\prod_{j=n_1}^{n} (1 - a_j) < \varepsilon$ for all $n > n_2$. Then, since products with indices exceeding N increase with fewer terms,

$$\begin{aligned}&\max_{N \leq k \leq n} \left|\frac{P_n}{P_{k-1}} (s_k - s)\right| \\ &\quad = \max\left\{\max_{N \leq k \leq n_1} \prod_{j=k}^{n} (1 - a_j)|s_k - s|, \max_{N \leq k \leq n} \prod_{j=k}^{n} (1 - a_j)|s_k - s|\right\} \\ &\quad \leq \max\left\{\max_{N \leq k \leq n_1} |s_k - s|\varepsilon, (1 - a_n)\varepsilon\right\} = \text{const } \varepsilon.\end{aligned}$$

Setting

$$a_{nj} = a_j \prod_{i=j+1}^{n} (1 - a_i),$$

we have for the maximum of the final term in Equation L2.1,

$$\max_{N \leq k \leq n} \left| \sum_{j=k}^{n} a_{nj}(s_j - s) \right| \leq \sum_{j=N}^{n} a_{nj}|s_j - s|. \tag{L2.2}$$

We see that $a_{nj} \to 0$ as $n \to \infty$ for each fixed j and, using Lemma 1, we find that

$$\sum_{j=N}^{n} a_{nj} = 1 - \frac{P_n}{P_{N-1}} \to 1$$

as $n \to \infty$. From $|s_j - s| \to 0$ as $j \to \infty$ and the Toeplitz Lemma (Knopp, 1947, p. 75) we infer that the bound in Equation L2.2 must go to zero as $n \to \infty$. Q.E.D.

Lemma 3

Let $\{\alpha_n\}$ and $\{\sigma_n\}$ be positive number sequences such that $\alpha_n \leq 2$ for all $n \geq N$ and $\sum \sigma_n < \infty$. If

$$e_{n+1}^2 \leq (1 - \alpha_n)^2 e_n{}^2 + \sigma_n(1 + e_n),$$

then

$$\sup_n e_n{}^2 < \infty.$$

It is *a fortiori* true that the conclusion is valid when $(1 - \alpha_n)^2$ is replaced by the smaller quantity $1 - 2\alpha_n$.

Proof. If $e_n{}^2 \leq 1$, then clearly $e_{n+1}^2 \leq M < \infty$ for any n. If $e_n{}^2 > 1$, then $e_n < e_n{}^2$ and

$$e_{n+1}^2 \leq (1 + \sigma_n)e_n{}^2 + \sigma_n$$

if $n \geq N$, because $(1 - \alpha_n)^2 + \sigma_n \leq 1 + \sigma_n$ when $\alpha_n \leq 2$. In every case, therefore,

$$e_{n+1}^2 \leq \max\{M, (1 + \sigma_n)e_n{}^2 + \sigma_n\} \qquad (n \geq N).$$

If we iterate this back to N, we find that

$$e_{n+1}^2 \leq \max\left\{\max_{N \leq k \leq n}\left[\frac{MP_n}{P_k} + \sum_{j=k+1}^{n} \frac{P_n}{P_j}\sigma_j\right], \frac{P_n}{P_{N-1}} e_N{}^2 + \prod_{j=N}^{n} \frac{P_n}{P_j}\sigma_j\right\}, \tag{L3.1}$$

where

$$P_n = \prod_{j=1}^{n} (1 + \sigma_j).$$

Since $1 + x \leq e^x$ is valid for all real numbers, we have

$$\frac{P_n}{P_k} \leq \exp \sum_{j=k+1}^{n} \sigma_j \leq \exp \sum_{j=1}^{n} \sigma_j < \infty,$$

for all $N \leq k \leq n$ and all $n \geq N$. The assumed summability of $\{\sigma_n\}$ thus shows that both terms in Equation L3.1 must remain uniformly bounded. Q.E.D.

Lemma 4

Let $\{b_n\}$ be any real-number sequence such that $\sum \beta_n^2 < \infty$, where $\beta_n = b_n^2/B_n^2$ and $B_n^2 = b_1^2 + \cdots + b_n^2$ (that is, Assumption A5‴). Suppose $z > 0$ and $K \geq 0$. Define

$$z_n = z\beta_n - K\beta_n^2,$$

and let N be fixed sufficiently large so that $0 < z_n < 1$ holds for all $n \geq N$. Then

$$C_{k-1}\frac{B_{k-1}^{2z}}{B_n^{2z}} \leq \prod_{j=k}^{n}(1 - z_j) \leq D_{k-1}\frac{B_{k-1}^{2z}}{B_n^{2z}}, \qquad k = N, N+1, \cdots, n$$

for some $0 < C_{k-1} \leq 1 \leq D_{k-1} < \infty$, which do not depend on n and have the property

$$\lim_{k\to\infty} C_k = \lim_{k\to\infty} D_k = 1.$$

Proof. The left-hand inequality in

$$\exp\left\{-\frac{x}{1-x}\right\} < 1 - x < \exp\{-x\} \tag{L4.1}$$

is valid for all $x < 1$. (See Knopp, 1947, p. 198, for this as well as Equation L4.4.) We set $x = z_j$ and form the product on j from k to n. Since

$$\sum_{j=k}^{n}\frac{z_j}{1-z_j} = \sum_{j=k}^{n}\left(z_j + \frac{z_j^2}{1-z_j}\right) \leq \sum_{j=k}^{n} z_j + \sum_{j=k}^{\infty}\frac{z_j^2}{1-z_j},$$

this gives

$$C_{k-1}\exp\left\{-\sum_{j=k}^{n} z_j\right\} \leq \prod_{j=k}^{n}(1 - z_j) \leq \exp\left\{-\sum_{j=k}^{n} z_j\right\}, \tag{L4.2}$$

with

$$C_{k-1} = \exp\left\{-\sum_{j=k}^{\infty}\frac{z_j^2}{1-z_j}\right\}$$

tending to 1 as $k \to \infty$ because $\{z_j^2\}$ is summable. From the right-hand inequality in Equation L4.1, we have

$$\frac{B_{j-1}^2}{B_j^2} = 1 - \beta_j < \exp\{-\beta_j\};$$

therefore,

$$\frac{B_{k-1}^2}{B_n^{\,2}} = \prod_{j=k}^{n} \frac{B_{j-1}^2}{B_j^{\,2}} \leq \exp\left\{-\sum_{j=k}^{n} \beta_j\right\}. \tag{L4.3}$$

But $z_j \leq z\beta_j$, and therefore

$$\frac{B_{k-1}^{2z}}{B_n^{\,2z}} \leq \exp\left\{-\sum_{j=k}^{n} z_j\right\}.$$

This combines with Equation L4.2 to give the asserted lower bound.

To prove the upper bound, we use

$$\exp\{x\} > \left(1 + \frac{x}{y}\right)^y, \tag{L4.4}$$

which is valid for all positive numbers x and y. For the choices

$$x = z_j, \qquad \text{and} \qquad y = \frac{B_{j-1}^2}{b_j^{\,2}}\, z_j,$$

we find that

$$\begin{aligned}
\exp\{z_j\} &\geq \left(\frac{B_j^{\,2}}{B_{j-1}^2}\right)^{(1-\beta_j)(z-K\beta_j)} \\
&= \left(\frac{B_j^{\,2}}{B_{j-1}^2}\right)^{z}\left(\frac{1}{1-\beta_j}\right)^{-(z+K)\beta_j + K\beta_j^{\,2}} \\
&\geq \left(\frac{B_j^{\,2}}{B_{j-1}^2}\right)^{z}\left(\frac{1}{1-\beta_j}\right)^{-(z+K)\beta_j}
\end{aligned}$$

because $1/(1-\beta_j)$ exceeds 1. Consequently, after inverting and forming the product over j, we have

$$\exp\left\{-\sum_{j=k}^{n} z_j\right\} \leq \prod_{j=k}^{n} \frac{B_{j-1}^{2z}}{B_j^{\,2z}}\left(\frac{1}{1-\beta_j}\right)^{(z+K)\beta_j} = \frac{B_{k-1}^{2z}}{B_n^{\,2z}}\, S, \tag{L4.5}$$

where

$$0 < \log S = (z + K)\sum_{j=k}^{n} \beta_j \log\frac{1}{1-\beta_j} \leq (z + K)\sum_{j=k}^{\infty} \beta_j \log\frac{1}{1-\beta_j}. \tag{L4.6}$$

Equations L4.5 and L4.6 combine with Equation L4.2 to give the asserted upper bound with

$$D_{k-1} = \exp\left\{(z + K)\sum_{j=k}^{\infty} \beta_j \log\frac{1}{1-\beta_j}\right\}.$$

Setting $x = \beta_j$ in the left-hand member of Equation L4.1, we see that the last written sum is majorized by $\sum_{j=k}^{\infty} \beta_j^{\,2}/(1-\beta_j)$. This goes to 0 as $k \to \infty$ since it is the tail of a convergent series, and therefore $D_k \to 1$. Q.E.D.

Lemma 5

Let $\{b_n\}$ be any real-number sequence such that

$$B_n^2 = b_1^2 + \cdots + b_n^2 \to \infty \quad \text{and} \quad \beta_n = b_n^2/B_n^2 \to 0 \text{ as } n \to \infty$$

(that is, Assumptions A3 and A5″). Define

$$\beta_{nk}(z) = \frac{B_k^{2z}\beta_k}{B_n^{2z}} \qquad (k = 1, 2, \cdots, n)$$

for $z > 0$. Then

$$\lim_n \sum_{k=1}^{n} \beta_{nk}(z)\xi_k = \frac{\xi}{z}$$

if $\lim_n \xi_n = \xi$, finite or not.

Proof. For every fixed k, $\beta_{nk} \to 0$ as $n \to \infty$. The conclusion follows immediately from the Toeplitz Lemma (Knopp, 1947, p. 75) if we can show that the row sums

$$R_n = \frac{\sum_{k=1}^{n} B_k^{2z}\beta_k}{B_n^{2z}} \to \frac{1}{z}$$

as $n \to \infty$. By the Abel–Dini Theorem, Equation 2.27, the numerator as well as the denominator approaches $+\infty$ with n. However, the value of $\lim R_n$ is obtainable as the limiting value of the ratio of successive numerator differences to denominator differences (Hobson, 1957, p. 7), that is, of

$$\frac{B_{n+1}^{2z}\beta_{n+1}}{B_{n+1}^{2z} - B_n^{2z}} = \frac{\beta_{n+1}}{1 - (1 - \beta_{n+1})^z}.$$

This ratio, in turn, is indeterminate (0/0), as $\beta_n \to 0$. But we can replace β_n by a continuous variable β and apply L'Hospital's rule to the resulting function. Thus, after differentiating, we have

$$\lim_n R_n = \lim_{\beta \to 0} \frac{1}{z(1 - \beta)^{z-1}} = \frac{1}{z}. \qquad \text{Q.E.D.}$$

Lemma 6

Let $\{b_n\}$ be any real-number sequence such that

$$B_n^2 = b_1^2 + \cdots + b_n^2 \to \infty \quad \text{and} \quad \sum \beta_n^2 < \infty,$$

where $\beta_n = b_n^2/B_n^2$ (that is, Assumptions A3 and A5‴). Then, for any $z > \frac{1}{2}$,

$$\Psi_n^2(z) = B_n^2 \sum_{k=N}^{n} \prod_{j=k+1}^{n} (1 - z\beta_j)^2 \frac{\beta_k}{B_k^2} \to \frac{1}{2z - 1}$$

as $n \to \infty$, where N is chosen so that $z\beta_j < 1$ for all $j > N$.

Proof. From Lemma 4, with $K = 0$, we have

$$C_k^2 \frac{B_k^{4z}}{B_n^{4z}} \le \prod_{j=k+1}^{n} (1 - z\beta_j)^2 \le D_k^2 \frac{B_k^{4z}}{B_n^{4z}},$$

where C_k^2 and D_k^2 both tend to 1 as $k \to \infty$. We thus have

$$\sum_{k=N}^{n} \beta_{nk}(2z - 1)C_{k+1}^2 \le \Psi_n^2(z) \le \sum_{k=N}^{n} \beta_{nk}(2z - 1)D_{k+1}^2,$$

where $\beta_{nk}(\cdot)$ was defined in the hypothesis of Lemma 5 for all positive arguments. After we take limits on both sides of this inequality, we find that the desired conclusion follows from that of Lemma 5. Q.E.D.

Lemma 7

(*a*) Let $\mathbf{B}$ be a positive definite $p \times p$ matrix with eigenvalues $0 < \lambda_1 \le \lambda_2 \le \cdots \le \lambda_p$ and associated unit eigenvectors $\boldsymbol{\phi}_1, \boldsymbol{\phi}_2, \cdots, \boldsymbol{\phi}_p$. Then, for every vector $\mathbf{x}$,

$$\frac{\mathbf{x}'\mathbf{B}\mathbf{x}}{\|\mathbf{x}\| \ \|\mathbf{B}\mathbf{x}\|} \ge \frac{2\sqrt{\lambda_p/\lambda_1}}{1 + \lambda_p/\lambda_1},$$

with equality holding if

$$\mathbf{x} = k\left[\left(\frac{\boldsymbol{\phi}_1}{\sqrt{\lambda_1}}\right) + \left(\frac{\boldsymbol{\phi}_p}{\sqrt{\lambda_p}}\right)\right].$$

(*b*) Let $\{\mathbf{h}_n\}$ be a sequence of p-dimensional vectors satisfying Assumptions F2, F3, F4, and F5. Then

$$\limsup_{n \to \infty} \frac{\lambda_{\max}\left(\sum_{j=1}^{n} \mathbf{h}_j\mathbf{h}_j'\right)}{\lambda_{\min}\left(\sum_{j=1}^{n} \mathbf{h}_j\mathbf{h}_j'\right)} \le \frac{K^{2q}}{\tau^2},$$

where

$$K = \max\left[\limsup_{n \to \infty} \|\mathbf{h}_n\|/\|\mathbf{h}_{n+1}\|, \limsup_{n \to \infty} \|\mathbf{h}_n\|/\|\mathbf{h}_{n-1}\|\right],$$

and

$$\tau^2 = \liminf_{k \to \infty} \frac{1}{p_k}\left[\lambda_{\min} \sum_{j \in J_k} \mathbf{h}_j\mathbf{h}_j'/\|\mathbf{h}_j\|^2\right].$$

Proof of a. Let $\mathbf{y} = \mathbf{B}^{1/2}\mathbf{x}$. Then

$$\frac{\mathbf{x}'\mathbf{B}\mathbf{x}}{\|\mathbf{x}\|\,\|\mathbf{B}\mathbf{x}\|} = \frac{\mathbf{y}'\mathbf{y}}{\|\mathbf{B}^{-1/2}\mathbf{y}\|\,\|\mathbf{B}^{1/2}\mathbf{y}\|} = \frac{\mathbf{y}'\mathbf{y}}{\sqrt{(\mathbf{y}'\mathbf{B}^{-1}\mathbf{y})(\mathbf{y}'\mathbf{B}\mathbf{y})}}.$$

The last is bounded below by $2\sqrt{\lambda_p/\lambda_1}/(1 + \lambda_p/\lambda_1)$ according to Theorem 13 in Beckenback and Bellman (1961). The second statement is verified by substitution.

Proof of b.

$$\lambda_{\max}\left(\sum_{j=1}^{n} \mathbf{h}_j\mathbf{h}_j'\right) \le \operatorname{tr}\left(\sum_{j=1}^{n} \mathbf{h}_j\mathbf{h}_j'\right) = \sum_{j=1}^{n} \|\mathbf{h}_j\|^2 \le \sum_{j=1}^{k+1} \sum_{i \in J_j} \|\mathbf{h}_i\|^2.$$

$$\lambda_{\min}\left(\sum_{j=1}^{n} \mathbf{h}_j\mathbf{h}_j'\right) > \lambda_{\min}\left(\sum_{j=1}^{\nu_k} \mathbf{h}_j\mathbf{h}_j'\right) \ge \sum_{j=1}^{k-1} \lambda_{\min}\left(\sum_{i \in J_j} \mathbf{h}_i\mathbf{h}_i'\right),$$

where k is chosen to be the largest integer for which $\nu_k \le n$. But,

$$\lambda_{\min}\left(\sum_{i \in J_j} \mathbf{h}_i\mathbf{h}_i'\right) \ge \min_{i \in J_j} \|\mathbf{h}_i\|^2 \tau_j^2 p_j,$$

where

$$\tau_j^2 = \frac{1}{p_j}\lambda_{\min}\left(\sum_{i \in J_j} \frac{\mathbf{h}_i\mathbf{h}_i'}{\|\mathbf{h}_i\|^2}\right)$$

and, of course,

$$p_j = \nu_{j+1} - \nu_j, \text{ the number of elements in } J_j.$$

Let

$$K_j = \max_{i,k \in J_j} \frac{\|\mathbf{h}_i\|}{\|\mathbf{h}_k\|}.$$

Then

$$p_j \min_{i \in J_j} \|\mathbf{h}_i\| \ge K_j^{-2} p_j \max_{i \in J_j} \|\mathbf{h}_i\|^2 \ge \frac{1}{K_j^2} \sum_{i \in J_j} \|\mathbf{h}_i\|^2,$$

and therefore,

$$\lambda_{\min}\left(\sum_{i \in J_j} \mathbf{h}_i\mathbf{h}_i'\right) \ge \left(\frac{\tau_j^2}{K_j^2}\right) \sum_{i \in J_j} \|\mathbf{h}_i\|^2.$$

Thus, we have

$$\frac{\lambda_{\max}\left(\sum_{j=1}^{n} \mathbf{h}_j\mathbf{h}_j'\right)}{\lambda_{\min}\left(\sum_{j=1}^{n} \mathbf{h}_j\mathbf{h}_j'\right)} \le \frac{\sum_{j=1}^{k} \sum_{i \in J_j} \|\mathbf{h}_i\|^2}{\sum_{j=1}^{k} (\tau_j^2/K_j^2) \sum_{i \in J_j} \|\mathbf{h}_i\|^2} + \frac{\sum_{i \in J_{k+1}} \|\mathbf{h}_i\|^2}{\sum_{j=1}^{k} (\tau_j^2/K_j^2) \sum_{i \in J_j} \|\mathbf{h}_i\|^2}.$$

The numerator of the second term is bounded by

$$K^{2(q+1)} \|\mathbf{h}_{\nu_{k+1}-1}\|^2 q.$$

Since $\liminf_{j\to\infty} \tau_j^2/K_j^2 \geq \tau^2/K^{2q}$, Assumption F2 implies that the second term approaches zero as $k \to \infty$. The first term on the right-hand side is indeterminate. The discrete version of L'Hospital's rule can be applied (Hobson, 1957, p. 7, Section 6), and we find that

$$\limsup_{k\to\infty} \sum_{j=1}^{k} \sum_{i\in J_j} \|\mathbf{h}_i\|^2 \Big/ \sum_{j=1}^{k} (\tau_j^2/K_j^2) \sum_{i\in J_j} \|\mathbf{h}_i\|^2 \leq \limsup_{j\to\infty} K_j^2/\tau_j^2 \leq K^{2q}/\tau^2.$$

This is the desired result. Q.E.D.

Lemma 8

Let $r_1, r_2, \cdots, r_n$ be any distinct real numbers, and let $\mathrm{R} = [\mathbf{r}_1, \mathbf{r}_2, \cdots, \mathbf{r}_n]$ be the (Vandermonde) matrix whose jth column is

$$\mathbf{r}_j = \begin{bmatrix} 1 \\ r_j \\ r_j^2 \\ \vdots \\ r_j^{n-1} \end{bmatrix}$$

Then the ith row of R^{-1} is

$$\frac{1}{d_i} [g_{n-1}(r_i), \cdots, g_2(r_i), g_1(r_i), 1],$$

where

$$d_i = \prod_{k\neq i} (r_i - r_k),$$

$$g_k(x) = x^k - (a_1x^{k-1} + a_2x^{k-2} + \cdots + a_k) \qquad (k = 1, 2, \cdots, n),$$

and $a_1, \cdots, a_n$ are such that

$$\prod_{k=1}^{n} (x - r_k) = g_n(x).$$

Proof. We introduce the $n \times n$ matrix

$$\mathrm{A} = \left[\begin{array}{c|cccc} 0 & & & & \\ \vdots & & \mathrm{I} & & \\ 0 & & & & \\ \hline a_n & a_{n-1} & \cdots & a_2 & a_1 \end{array}\right],$$

and notice that $\mathbf{A}\mathbf{r}_j = r_j\mathbf{r}_j$ because $x = r_j$ is a root of $g_n(x) = 0$. In other words, r_j is the eigenvalue of $\mathbf{A}$ associated with the right-sided eigenvector $\mathbf{r}_j$. If we post-multiply $\mathbf{A} - x\mathbf{I}$ by

$$\mathbf{E}(x) = \begin{bmatrix} 1 & 0 & 0 & \cdots & 0 \\ x & 1 & 0 & \cdots & 0 \\ x^2 & x & 1 & \cdots & 0 \\ \vdots & \vdots & \vdots & & \vdots \\ x^{n-1} & x^{n-2} & x^{n-3} & \cdots & 1 \end{bmatrix}$$

we find that

$$(\mathbf{A} - x\mathbf{I})\mathbf{E}(x) = \left[\begin{array}{c|ccc} 0 & & & \\ \vdots & & \mathbf{I} & \\ 0 & & & \\ \hline -g_n(x) & -g_{n-1}(x) & \cdots & -g_1(x) \end{array}\right]$$

identically in all real numbers x. Now set

$$\mathbf{l}_i' = [g_{n-1}(r_i), \cdots, g_1(r_i), 1].$$

Then

$$\mathbf{l}_i'(\mathbf{A} - r_i\mathbf{I})\mathbf{E}(r_i) = [0, 0, \cdots, 0].$$

But $\mathbf{E}(x)$ is nonsingular for all x (its determinant is unity), so $\mathbf{l}_i$ must be a left-sided eigenvector of $\mathbf{A}$. Letting $\mathbf{L}$ be the matrix whose columns are these vectors, we can easily see that $\mathbf{L}'\mathbf{R}$ commutes with the diagonal matrix of roots $r_1, r_2, \cdots, r_n$; in terms of entries, $(r_i - r_j)\mathbf{l}_i'\mathbf{r}_j = 0$. By hypothesis, the r's are distinct numbers, and therefore the two sets of eigenvectors are biorthogonal. In other words, we have

$$\mathbf{L}'\mathbf{R} = \mathbf{D}, \qquad \text{or} \qquad \mathbf{R}^{-1} = \mathbf{D}^{-1}\mathbf{L}'$$

for some diagonal matrix $\mathbf{D}$.

We complete the proof by showing that the ith entry of $\mathbf{D}$ is indeed the one given in the statement of the lemma. To do this, we multiply $g_k(x)$ by x^{n-k-1} and sum on k up to $n - 1$, giving

$$\sum_{k=1}^{n-1} g_k(x)x^{n-k-1} = (n - 1)x^{n-1} - \sum_{k=1}^{n-1}\sum_{j=1}^{k} a_j x^{n-j-1}.$$

After collecting the coefficients of the a's, we have

$$\begin{aligned} x^{n-1} + \sum_{k=1}^{n-1} g_k(x)x^{n-k-1} &= nx^{n-1} - \sum_{k=1}^{n-1} a_k(n - k)x^{n-k-1} \\ &= \frac{d}{dx}\left[x^n - \sum_{k=1}^{n-1} a_k x^{n-k} + C\right]. \end{aligned}$$

If we set the arbitrary constant equal to $-a_n$, the right-hand side is the derivative of $g_n(x)$. After differentiating the product form of the characteristic polynomial, we obtain the identity

$$x^{n-1} + \sum_{k=1}^{n-1} g_k(x)x^{n-k-1} = \sum_{k=1}^{n} \prod_{j \neq k} (x - r_j).$$

In particular, at $x = r_i$ the left-hand side becomes the inner product, $\mathbf{l}_i'\mathbf{r}_i$, and the right-hand side becomes $\prod_{j \neq i} (r_i - r_j)$. Q.E.D.

The reader will note that the preceding is precisely the sort of analysis used in deriving the solution of an nth-order difference equation with coefficients $a_1, \cdots, a_n$. The proof immediately suggests itself if one knows that the right-sided eigenvectors of the matrix defining the corresponding first-order vector difference equation is a Vandermonde matrix of characteristic roots. Only the point of view is different; we start with the characteristic roots and construct the coefficients.

References

Beckenback, E. F., and R. Bellman (1961). *Inequalities*, Springer-Verlag, Berlin.

Burkholder, D. L. (1956). "On a Class of Stochastic Approximation Processes," *Ann. Math. Statist.*, **27**, 1044–1059.

Chernoff, H. (1956). "Large Sample Theory: Parametric Case," *Ann. Math. Statist.*, **25**, 463–483.

Chernoff, H. (1959). "The Sequential Design of Experiments," *Ann. Math. Statist.*, **30**, 755–770.

Hobson, E. W. (1957). *The Theory of Functions of a Real Variable and the Theory of Fourier Series*, Vol. 2, Dover, New York.

Hodges, J. L., Jr., and E. L. Lehmann (1951). "Some Applications of the Cramér-Rao Inequality," *Proceedings of the Second Berkeley Symposium on Mathematical Statistics and Probability*, University of California Press, Berkeley, California, pp. 13–22.

Hodges, J. L., Jr., and E. L. Lehmann (1956). "Two Approximations to the Robbins–Monro Process," *Proceedings of the Third Berkeley Symposium on Mathematical Statistics and Probability*, Vol. 1, University of California Press, Berkeley, California, pp. 96–104.

Householder, A. S. (1964). *The Theory of Matrices in Numerical Analysis*, Blaisdell, New York.

Kalman, R. E. (1960). "A New Approach to Linear Filtering and Prediction Problems," *Journal of Basic Engineering*, **82**, 35–45.

Knopp, K. (1947). *Theory and Application of Infinite Series*, Hafner, New York.

LeCam, L. (1953). *On Some Asymptotic Properties of Maximum Likelihood Estimates and Related Bayes Estimates*, University of California Publications in Statistics, **1**, No. 11, 277–330.

Loève, M. (1960). *Probability Theory*, Van Nostrand, New York.

Parzen, E. (1960). *Modern Probability Theory and its Applications*, Wiley, New York.

Robbins, H., and S. Monro (1951). "A Stochastic Approximation Method," *Ann. Math. Statist.*, **22**, 400–407.

Sacks, J. (1958). "Asymptotic Distribution of Stochastic Approximation Procedures," *Ann. Math. Statist.*, **29**, 373–405.

Index

Frequently Used Symbols, and Where Defined or Redefined

www.ingramcontent.com/pod-product-compliance
Lightning Source LLC
Chambersburg PA
CBHW061211220326
41599CB00025B/4606